User-Centered Information Design for Improved Software Usability

For a complete listing of the *Artech House Computer Science Library,*
turn to the back of this book.

User-Centered Information Design for Improved Software Usability

Pradeep Henry

AH

Artech House
Boston • London

Library of Congress Cataloging-in-Publication Data
Henry, Pradeep.
    User-centered information design for improved software usability /
    Pradeep Henry.
       p.   cm.
    Includes bibliographical references and index.
    ISBN 0-89006-946-8 (alk. paper)
    1. User interfaces (Computer systems)   2. Application software—
Development   I. Title
QA76.9.U83H46   1998
005.4'28—dc21   98-6544
                                                                  CIP

British Library Cataloguing in Publication Data
Henry, Pradeep.
    User-centered information design for improved software usability
    1. Human-computer interaction
    I. Title
    004'.019

    ISBN 0-89006-946-8

Cover design by Elaine K. Donnelly

© 1998 ARTECH HOUSE, INC.
685 Canton Street
Norwood, MA 02062

International Standard Book Number: 0-89006-946-8
Library of Congress Catalog Card Number: 98-6544

10 9 8 7 6 5 4 3 2 1

To the memory of Della,
an unassuming achiever who believed I could achieve anything.

Contents

Preface

WELCOME. You are holding a book on enhancing software usability. This book is about designing the four information components of software usability: labels, messages, online support elements such as Help, and printed support elements such as a user's guide. You will soon discover how to apply an integrated approach to designing these information components.

This is the first book to approach technical writing—all of it, including the design of labels and messages—from its role in the larger software usability context. The book's software usability-driven approach to information design is called *user-centered information design* (UCID). In UCID, the design activity is iterative and evaluation is a two-level approach. Technical writers, in collaboration with usability engineers and software engineers, primarily design and write all the information components, including labels and messages.

Benefits of implementing UCID

By giving you a software usability-driven approach, this book puts you on the right track toward designing information. UCID is important to you because it benefits products, users, and software organizations. By helping you meet users' information requirements, UCID achieves improved software usability, which impacts your organization's image and the bottom line.

This book concisely presents pioneering strategies for the design and evaluation of information. The critical design goals you set and the two-level evaluation approach keep your information development efforts focused on improving software usability.

You cannot plan and build a bridge with a single deck, and later attach another deck to it. Integration is key to UCID, too. The UCID process ensures that individual information elements (Help, user's guide, etc.) are identified and designed together as one piece—with improved software usability as the goal. You can now make better decisions on what types of information to provide in which media and in what form—based on users' and project-specific requirements. Through integration, you will avoid four undesirable things: exclusion of critical information, inclusion of unnecessary information, unnecessary repetition, and information inconsistencies across the software system.

A UCID project makes full use of the technical writer's skills. Writers have an expanded and challenging role that includes the design of labels and messages.

Where did UCID come from?

If your goal is to design information that maximizes software usability, you'll find that traditional technical writing approaches and processes are not entirely adequate. During my adventures described below, I realized the existing inadequacies and the need for a software usability-driven approach.

- *Project experience.* I've learned my best lessons while working on a large software development project in which the client wanted the product to be high on usability, but we had no professional usability specialists or technical writers. Before we could finally ensure client satisfaction, we had to change our processes, and I had to contribute my technical writing skills to the design of labels, messages, and online Help, besides printed manuals.

- *Classroom interaction*. I've learned from interactions with participants at corporate courses I've taught in technical writing and user interface design. I've also learned from the experiences shared by a creative bunch of silicon valley professionals at a UC Berkeley (Extension) usability course I took recently.

- *Literature research*. Usability professionals have always believed that the four information components impact software usability. From their papers, books, and compilations of usability principles, I've learned about each information component's role in improving software usability.

These adventures and project success stories originated UCID as the answer to the need for a software usability-driven approach to information design.

How this book is organized

This book has 14 chapters. The first chapter helps you see the product usability context under which information appears, so that you can better understand the role of information. The next three chapters lay the foundation for UCID. Chapters 5 through 8 guide you in the crucial preliminary activities of analysis, goal-setting, high-level design, and low-level design. The next four chapters, chapters 9 through 12, focus on the four information components. Chapter 13 lists information quality goals and describes design techniques to achieve those goals. The last chapter is about evaluating information. Finally, the glossary defines all the new terms, such as "critical design goals," and terms with special meaning, such as "end-use task."

Specifically, here's the purpose and content of each chapter.

Chapter 1 lays the conceptual foundation for the remainder of the book. Toward helping you better understand the role of information, this chapter looks at how people use software, their needs, and how they respond when their needs are not satisfied. Also, this chapter identifies and categorizes all the things that impact software usability. You will see how information contributes to software usability.

Chapter 2 introduces UCID and describes the key features of this software usability-driven approach. You will learn what UCID can do for you and how you can integrate it into your current software engineering process.

Chapter 3 starts with a brief description of traditional processes for information development. The user-centered design process, under which UCID should appear, is discussed. Following that is a description of the key features of the UCID process, showing how it overcomes the limitations of current processes. Finally, we go into specific UCID activities.

There are certain things that make a UCID project unique. For example, new approaches are introduced and technical writers are now empowered with expanded roles. This calls for a focus on certain management related activities. Chapter 4 first introduces the key contributors of UCID and then goes on to describe these activities: managing collaboration and change, preparing UCID and software usability plans, and implementing a metrics program.

To be effective, all software usability components must be designed around tasks that users will perform—while considering the users' knowledge, capabilities, limitations, and environment. The activity of getting to know users and tasks is described in Chapter 5.

There are two objectives for Chapter 6. One is to define software usability goals through interdisciplinary collaboration. The other is to define information quality goals that contribute to maximizing the software usability goals. Toward defining information quality goals, we will learn how information is used by users. We will also learn how to use the "information use model" to come up with a set of *critical design goals* (CDGs), which are factors critical for the success of each information element, such as a user's guide, you may provide.

Chapter 7 is about determining the right combination of information elements that will maximize the usability of the software system. The activity is called integration, and the resulting design is called information architecture.

Chapter 8 covers the two low-level design activities. First is the specification of the content and organization of each information element. Second is the creation of prototypes or models that represent each element (or a distinct section of it). The rest of the element will have to be written like the prototype.

Chapter 9 is about software labeling, which is a critical and difficult part of user-centered design. Technical writers can read this chapter before they attempt labeling.

Poor design of system feedback or messages can draw undesirable reactions from users, thereby impacting the usability and success of the software. Chapter 10 is about designing effective messages. Technical writers and programmers who coauthor messages should read this chapter.

In Chapter 11, we will look at some common support elements meant for the online medium. Technical writers can read this chapter to learn how to design Help, tutorials, and so forth.

Users often need detailed information on paper. In Chapter 12, we will look at support elements meant for the print media. Technical writers can read this chapter to learn how to design common printed manuals.

Chapter 13 provides techniques that you can use to achieve information design goals for various information elements. The complete list of information design goals (ARRCUPA) is covered. Technical writers can use this chapter as a mini-reference to learn various techniques to improve, say, retrievability.

Information is good quality only when it contributes to improved software usability. When this is your ultimate goal, you need an evaluation approach that will improve individual information elements so that they—in combination—achieve improved software usability. In addition, your approach should also cover labels and messages. Chapter 14 describes a two-level evaluation approach. You will learn how to use common testing, review, and editing methods to perform the two levels of evaluation.

Who can benefit from this book?

The primary audience for this book is comprised of technical writers and technical writing managers. Because this book is about designing usable software, it will interest other audiences as well. Usability engineers will find it interesting as information is an important software usability component. Because UCID is a process and requires interdisciplinary collaboration, the book will also be useful for software engineers, quality/process specialists, and project managers.

■ *Technical writers/managers.* If you are a technical writing manager, you will be able to show management how UCID can benefit the organization. You can also use this book as an overall planning guide to manage UCID projects. If you are a technical writer, the book provides you the training required for your expanded, more challenging role in UCID projects. You can also use it to guide you through each phase of the UCID process.

■ *Usability engineers/managers.* Labels, messages, online support elements (Help, etc.), and printed support elements (user's guides, etc.) are components of software usability. Windows and widgets are not all. This book provides the framework to help you successfully integrate UCID into your user-centered design process. It is also a useful guide to help you consider the information components in your goal-setting, design, and evaluation efforts. Further, it helps you knowledgeably manage the whole user-centered design effort including UCID.

■ *Software engineers.* You will learn what UCID is, and who does what in UCID projects. You will find guidelines and checklists for performing technical reviews. The chapter on designing messages will be useful for collaborating with technical writers to coauthor messages.

■ *Quality/process specialists.* Learn about the UCID process and how it can be integrated into your existing software engineering process. You will also learn about the information quality factors, which are essential when your goal is improved software usability.

■ *Project managers*. You can use it for overall planning and management of UCID projects.

The book is goal-driven and minimalist

This book is focused on designing information for improved software usability. I've maintained this focus via two ways. One is the goal-driven approach. The other is the minimalist approach.

Typically, communication books are organized around a document's "parts" such as preface, headings, and index. Readers often do not know why these parts should be provided in the first place. In contrast, almost everything in this book is goal-driven. At the top level, your ultimate goal is improved software usability, not "well-written" or "perfect" documents. Once you have identified various documents that need to be provided to users, you move to the second level, where you identify the goals for each document. These goals can be retrievability, accuracy, and so forth. Only then are you ready to look for techniques or "parts" that can help you achieve those goals. For example, if your goal is retrievability, you should explore parts such as a search tool and a good index. You can clearly see this approach in Chapter 13, where goals are listed and techniques described for each one.

The second technique I've used is minimalism. I have avoided details and topics that do not significantly contribute to improved software usability, and those that are not really helpful in understanding and implementing UCID. For example, this book does not address grammar or print production topics. Such material is available in other books.

I hope that this book's focused approach will also keep your information design efforts intent on improving software usability—your ultimate goal.

1

Information: Its Role in Software Usability

nformation is all the textual elements that software users see and use, including, but not limited to, system messages and printed manuals. And *software usability* is the ease with which users can use software.

This chapter sets up the conceptual foundation for the rest of the book. The objective is to help you see the product usability context under which information appears so that you can better understand the role of information. Toward that end, we will look at how people use software, their needs, and how they respond when these needs are not satisfied. We will also identify and categorize all the things that impact software usability and see how information contributes to it.

1.1 Software users: what they do, what they need

Why do people use software? Certainly, they use it to make life's tasks easier and quicker for themselves and for others, such as their customers. Therefore, one of our basic beliefs in this book is simply that people use software to perform tasks. Our goal then is to make it easy, quick, and pleasant for them to do so. To do this, we should know what tasks users perform and what knowledge they need to effectively perform those tasks.

1.1.1 Users perform tasks

What is a task? In e-mail software, composing the message to be sent is a task. Sending the message is another task. Viewing a message is also a task. These tasks are the purpose for which users or their employers bought the software. What is common in these tasks is that all of them are specific to the software. We'll call these *end-use* tasks.

To effectively use the software, users may have to perform certain other tasks besides end-use tasks. For example, before they can start using the software, users will have to perform installation. Such tasks may be required in most software to make end use possible. We'll call such common tasks *support tasks*. Here are some support tasks:

- *Installation.* Setting up a software system on the computer. The simplest installation task may only demand that the user copy files from the installation diskette to the hard disk. In contrast, installing software on a mainframe computer, for example, can be very complex, requiring the expertise of many systems specialists. Moreover, it may even take a few days to complete.

- *Customization.* Making changes to the software's "factory settings" to suit the user's or organization's needs. Often, built-in features are used for customization tasks. A simple example is changing the date format to match one required in the user's country.

- *Administration.* Managing computer resources to meet the organization's performance, security, and other requirements. This includes monitoring the use of the software to ensure that processing goals are met. It can also include charging users for their use of various computing resources.

■ *Operation.* Starting, controlling, and stopping various facilities or components of a software—typically at the system console. Operators issue commands that can affect the use of all the computing resources in the organization.

The significance of each of these support tasks depends on the features, use, and complexity of the software. If it is a large software package for the mainframe, systems specialists are likely to perform all of those tasks. In this case, all the support tasks may be substantial, each requiring its own manual. If it is a windowing networked software package, again systems specialists may perform the support tasks, not the people who perform end-use tasks. Here, the support task of *operation* may be absent. If it is a single-user personal computer software package, all the support tasks required may be performed by the end user. In this case, each support task may be described in separate chapters of the same manual. Figure 1.1 shows some end-use and support tasks.

"I'd be entering
customer details into
the accounting system."

"I'd be installing
and customizing
the new accounting system."

"I'd be performing
budgeting and forcasting tasks
with the accounting system."

End-use tasks | Support tasks

Figure 1.1   End-use and support tasks.

1.1.2   Users need to learn or recall specific knowledge

A software task, whether end-use or support, typically goes like this.

1. *Invoke the required task.* In today's personal computer software, this could be as simple as selecting a menu option.

2. *Enter data.* For example, enter amounts in an accounting software package or draw a circle in a graphics program.

3. *Receive results.* Results can be that which the user requested, or it could be system feedback such as a message.

4. *Respond to the results.* For example, correct an entry in response to an error message.

For this user-software interaction to happen effectively, users need to know certain things. The most important of these knowledge components are listed in Table 1.1.

At any given time, a user is unlikely to possess all the knowledge for all the tasks supported. Rather, a user will have a different mix at various stages of software use. In fact, the level of each knowledge component will also vary.

To perform a software task, a user needs to do one of two things: recall the knowledge if it is already known, or learn it if it is not known.

Table 1.1
Knowledge Components Users Need to Perform a Task

Application Domain Knowledge	Knowledge about the industry or field in which the software is used.
Computer-Use Knowledge	How to use the keyboard and mouse? How to use *this* operating system?
Task Concepts Knowledge	*Software-specific* concepts about a task. What is the task? Why perform? When to perform? Who should perform?
Task Procedure Knowledge	High-level steps involved in the task as required in *this* software.
User Interface Knowledge	How to navigate? How to select an option? What data to enter? How to respond to a message?

Recall if they already know Software systems should be designed to make it easy for users to recall the knowledge they already have about tasks. Recall depends on users' ability for retention, which in turn depends on things like the usability of the software system and the frequency of software use.

Learn if they do not know Usually, people must first learn how before they begin to regularly use a software system. Users may have moved to this software after using a competing product. Or, they may not have used this operating system before. Or, they may even be new to computers. Therefore, software systems should be designed to be quickly and easily learned.

The responsibility for the knowledge components listed in Table 1.1 is not entirely on users alone. In fact, except for application domain knowledge (and maybe computer-use knowledge), the responsibility is largely on the software house, or developer. Does that mean the software house should just write complete "documentation" and give it to users? Not at all. Rather, they should design software to help achieve the recall and learning objectives of users.

1.2 The wide gap between users and software

Often, software is not designed around tasks that users perform. Therefore, there is a wide gap between users and software. Users find themselves unable to comfortably or effectively use the software. See Figure 1.2.

1.2.1 Why does the gap exist?

The following are often reasons for the gap between users and software.

- *Slogan.* "It should work" is often the only slogan. Today, reliability or functionality alone is not enough. Usability needs to be high on the software house's priorities because it impacts users' productivity and overall satisfaction with the product.

- *Process.* The software development process follows an engineering model with little or no concern for the product's users. Users are

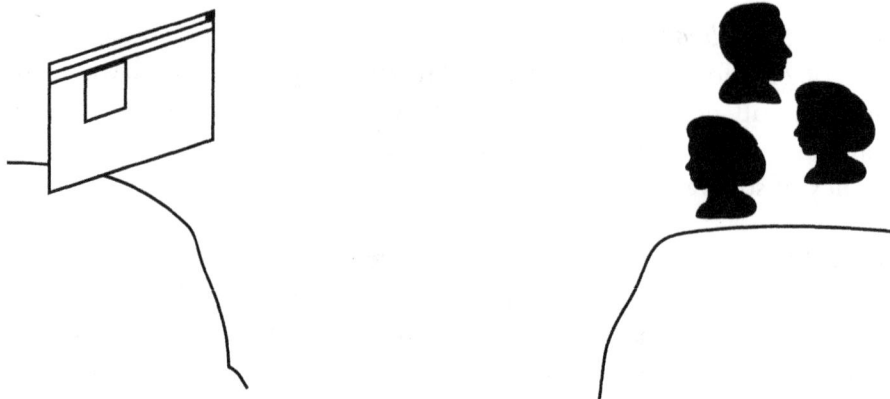

Figure 1.2 The gap between users and software.

neither understood nor involved in the design and evaluation of the software.

- *Approach.* First, system internals are designed. Then, a user interface is "slapped" on it. The resulting user interface reflects the underlying mechanism. However, for the user who just wants to complete a task using the software, a user interface based on the task is more appropriate than one based on system internals. The increasing numbers of users with little or no interest in the internal functions clearly shows the need for shielding them from the internals and providing user interfaces based on task models instead. For this, the interface should be designed first—based on users' tasks—and then the internals.

- *Skills.* Good software design calls for interdisciplinary skills. Besides programming skills, we need skills in user-centered design, technical writing, and graphic design.

1.2.2 How do users react?

In a *Tom and Jerry* comic, Tom returns home tired in the evening. He has to get up early the next day. For a wake-up alarm, Tom attempts to set his clock with the help of a user manual. Struggles. Gets a rooster instead and ties him up for a hassle-free wake-up call. (No manuals required here.) Tom did not give up the task. He gave up the hard-to-use product. And this is exactly what software houses today are worried about. People will switch to a more usable brand.

If the user-software gap is wide, users describe the software as unusable. Users' reactions to unusable software can range from the psychological to the physical. Here is an adaptation of the psychological responses reported in [1]:

- *Confusion.* The software is too complex. It is not structured as users expect.

- *Frustration.* The software is inflexible and unforgiving. Users cannot undo incorrect actions.

- *Panic.* The software's response time is too slow—just when the user is under tremendous pressure.

- *Boredom.* The software's response time is slow and tasks are overly simplified.

- *Incomplete use.* Only some of the software's functions or tasks are used. These tasks are often the easiest to perform.

- *Indirect use.* Managers who find the software hard to use, for example, get someone else to do it.

- *Misuse or modification.* Those who know the software well may change it to meet personal requirements that do not advance organizational interests. In this case, system integrity may be adversely impacted.

- *Abandonment.* The software is rejected by managers and other users who have the discretion to reject it.

1.3 The software usability bridge

We have seen that there is a wide gap between users and software. To eliminate the negative impact of this gap on the software organization, we need a bridge that will get users across. The good news is that software houses are already building bridges. "Across the country, software and hardware companies are building laboratories, testing products, hiring experts, contracting consultants, and rewriting code, all in the name of usability" [2]. Methods and techniques have been recommended to design software for usability. The process increasingly used today is called *user-centered design* or usability engineering.

1.3.1 What is software usability?

Some of the concerns for the acceptability of software are the software's utility (that is, whether or not it supports the end-use tasks users need), usability, reliability, support, and cost. Usability is a software's quality of being easy, quick, and pleasant to perform *end-use* and *support* tasks. Software systems with such characteristics actually promote learning and recall.

- The *easy* attribute. People want easy-to-use software. And that is nothing new—it is in keeping with the continuing trend toward making life easier, not harder.

- The *quick* attribute. Easy is often faster or more productive. But not always. People experienced with the software will need faster—not necessarily easier—ways of performing tasks. Response time and computing environment can impact user productivity.

- The *pleasant* attribute. This is a subjective, user-preference attribute. Things like good aesthetics and friendly tone can make it pleasant to use software.

In our definition of software usability, we have included a key point. Even support tasks such as installation should be easy, quick, and pleasant to perform. This is especially important in personal computer software, where support tasks are often performed by people who perform end-use tasks, people who need not be systems specialists.

Software that meets the requirements listed above satisfies users and customers. And that means improved image and bottom line for the software house that designed it.

1.3.2 Components of the software usability bridge

The characteristics of the following largely affect task performance or software usability:

1. *Users who will perform the tasks.* What do they already know? What do they need to learn?

2. *Use environment.* Is the software used in a noisy place? Is it a dusty place? What about the ergonomics of the environment, including furniture and lighting?

3. *Computing environment.* Are users' computers 486-based with 4 MB RAM or Pentium-based with 32 MB RAM?

4. *Interaction techniques used.* Is the user interface character-based or graphical? What is the style of interaction: is it form fill-in, direct manipulation, menu-driven, natural language, or command language? Should users first select an object then action, or the other way around?

5. *Information provided.* Do error messages mention corrective action? Do manuals give conceptual information?

Software houses have little or no control over the first three items above. However, they can and should design the last two—interaction and information—with a knowledge of users, use environment, and computing environment.

A closer look reveals further division into useful categorizations. *Interaction* can be divided into user objects and user actions. And *information* can be divided into interaction information and support information. We will soon define these terms. Until then, we can summarize the four major software usability components in Figure 1.3

Besides being an engineering marvel, the suspension bridge illustrates how individual components work together to help people get across to the other side. The four components of software usability can be thought of as the four major components of a suspension bridge: the towers, deck, cables, and suspenders. You see from Figure 1.4 how each component works with, for, and is necessary to the other—in order to achieve the best possible whole.

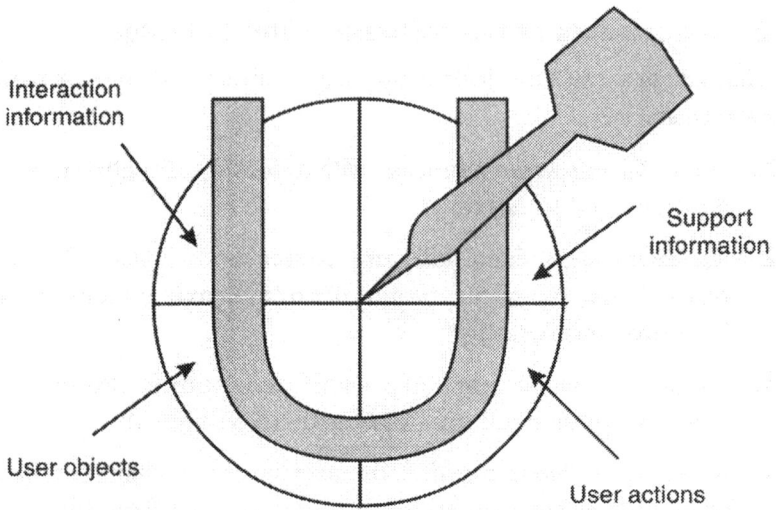

Figure 1.3   The four components of software usability.

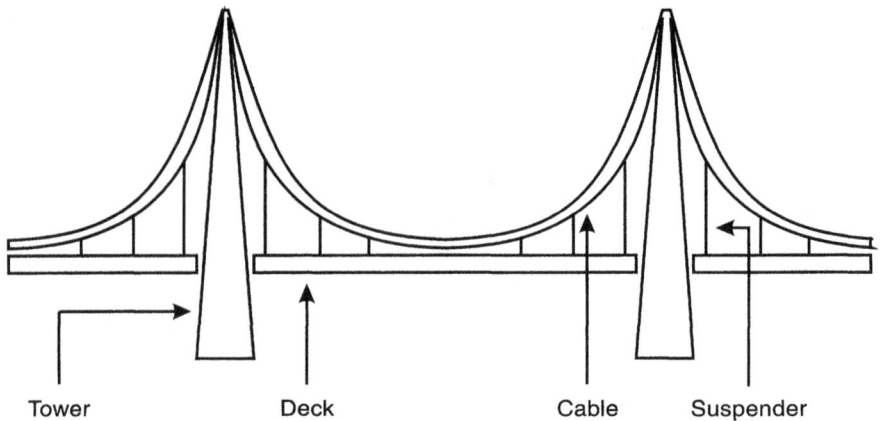

Figure 1.4   The suspension bridge: different parts working toward a single goal.

Our primary concern is to design software to improve the way people can perform tasks. Components are only tools used in the design. However, we have identified and categorized various software usability components because this knowledge helps us recognize every information component and assign tailored processes and resources for the development of each. The result will be overall improvement of product usability.

1.3.2.1 Interaction components

User objects and user actions are the things that primarily drive user-software interaction.

User objects User objects are the items users see (and interact with) on the software's screen. Examples are windows, menus, and scroll bars. Figure 1.5 gives a list of user objects.

User actions User actions are the interactions users have with user objects. An example is a mouse click a user performs to select an icon. User actions include the following:

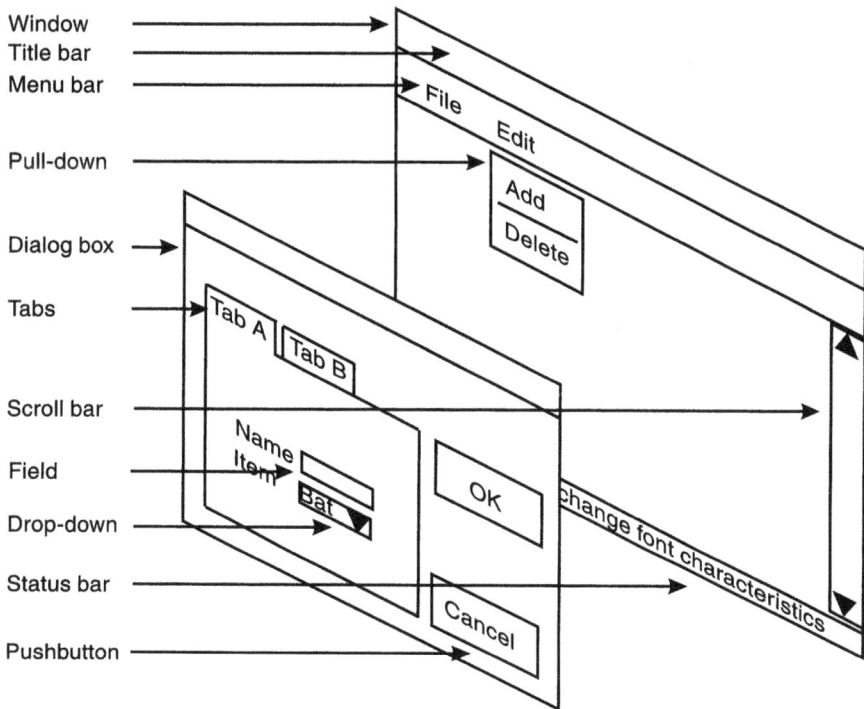

Figure 1.5 User objects.

- Navigation;
- Selection;
- Keystrokes (entry, etc.);
- Direct manipulation (clicking, dragging, dropping, etc.).

1.3.2.2   Information components

For effective use of the interaction components described above, users may need to use various pieces of textual information. This includes obvious ones like online Help and printed manuals, and less obvious ones like the labels used for identifying user objects. These pieces of information can be broadly grouped as interaction information and support information. This book is about the integrated design of interaction and support information.

Interaction information   Labels and messages are information essential for completing user-software interaction. Therefore, we will call them *interaction information*. Whereas labels are required to identify user objects such as menu options, messages are required to give feedback to users about a user action or a change in system status.

Support information   For effective use of a software system, we also need to provide information to "clarify" the following three software usability components: user objects, user actions, and interaction information. We will call this *support information*. More commonly called user documentation, support information is categorized as online support and printed support.

1.3.3   How to achieve software usability

For many decades, aircraft occasionally crashed because of "pilot error." Nuclear disasters, too, happened because of "human error." Then human factors specialists found that the disasters were actually due to poor user interface design, not human error. Human factors engineering is now a proven, interdisciplinary profession. *Human-computer interaction* (HCI) is the discipline that focuses on the usability of computer systems. Through research and study, HCI has come up with processes, standards, and tools for the design of usable computer systems. For the usability design of software systems, the process that is increasingly practiced today is called user-centered design. Many new methods, tools, and skills have been built around this process.

1.3.3.1 User-centered design

User-centered design (or usability engineering) is the approach customer-driven organizations are following today to design the user interface. It fits into popular flavors of the software engineering process. The user-centered design process is described in Chapter 3. In this section, we will look at key concepts common in a user-centered design process:

- *Focus early on users and tasks.* Understand users' cognitive, behavioral, and attitudinal characteristics, the tasks users perform, how, and in what kind of environment.

- *First design the user interface.* Separate user interface design from internal design. Reverse the traditional process by first designing the user interface.

- *Involve users.* Have users participate in design and design reviews.

- *Insist on iterative prototyping and evaluation.* Evolve user interface design via user testing and iteration.

1.3.3.2 Methods

There are many proven methods for user/task analysis, design, and evaluation. Design methods include prototyping and participatory design. Among the evaluation methods, we have usability inspection methods, such as heuristics evaluation and walk-throughs, and user testing methods such as laboratory and field tests.

1.3.3.3 Interdisciplinary skills

The creation of easy-to-use software involves a design team with members from various disciplines working together. The skill set for user-centered design practitioners includes the knowledge of HCI literature, cognitive processes, experimental design, rapid prototyping, quantitative methods, task analysis methods, observational techniques, usability testing, user interfaces, and HCI standards and guides [3]. Besides, a user-centered design process needs technical writing skills for designing Help and other information, and graphic design skills for user interface layout and icon design.

1.3.3.4   Software usability heuristics

Over the years, usability practitioners and researchers have come up with sets of principles of software usability. Here is a set of principles developed by Jakob Nielsen and Rolf Molich, and listed in [4]:

1. *Simple and natural dialog.* Dialogs should not have any irrelevant or infrequently used information. All information should be arranged in a way that is natural to users.

2. *Speak the user's language.* Dialogs should be expressed in text and concepts familiar to users.

3. *Minimize user memory load.* Users should not have to remember information as they move from one part of the dialog to another.

4. *Consistency.* Users should not have to wonder whether different words, situations, or actions mean the same thing.

5. *Feedback.* Users should always be informed about what is happening in the system.

6. *Clearly marked exits.* System should have visible exits so that users can leave any unwanted situation.

7. *Shortcuts.* Accelerators that speed up tasks should be available for expert users.

8. *Good error messages.* Messages should, in plain language, state the problem and suggest a solution.

9. *Prevent errors.* Systems should, whenever possible, prevent problems from occurring.

10. *Help and documentation.* Information should be easy to retrieve and should list required steps to complete tasks.

1.3.3.5   Standards and guidelines

Recently, activities relating to the development of user interface standards have been stepped up. Standards are being developed at three levels: the international level by organizations like the *International Organization for Standardization* (ISO), the national level by organizations like the *American National Standards Institute* (ANSI), and the vendor level by various windowing system developers. The vendor-level standards may be more important than the others because they specify the look and feel of the user interface in great detail. Examples are IBM's *Common User Access* (CUA) and Microsoft's Windows standards.

One reason why importance is given to standards is to ensure user interface consistency within and across software platforms. Consistency is believed to contribute to software usability as it helps users transfer their skills from one software system to another. Nielsen [4] reports that in several studies, consistency reduced user training time to between 25% and 50% of that needed for inconsistent user interfaces. Of course, there is a danger to standardization. It can inhibit creativity and innovation.

1.3.3.6   Tools

Perhaps the most common tools have been used for prototyping and laboratory testing.

Prototyping tools help designers rapidly construct a mock-up of the user interface for user testing. Some tools only let you create the visual components, such as screen layouts and dialog boxes. Others let you develop complete working prototypes.

Laboratory tools are primarily used for logging. With logging tools, you can record the time when events such as user actions occur. You can also annotate the events. The time stamps are usually coordinated with the timing codes on a videotape used in the lab test. Using a playback tool, you can retrieve the taped record of the test, starting a minute or so before each event to be studied. Besides logging tools, there are also ones for capturing keystrokes and other events during a lab test.

1.4 Information: critical component of the software usability bridge

Good interaction and support information play important roles in improving software usability. They support the learning and recall objectives of users.

1.4.1 How information contributes to software usability

Generally, when someone talks about software usability, he or she talks about the interaction components—that is, user objects and user actions. However, from an earlier discussion we know that software usability is also impacted by information components (interaction information and support information). Interaction information is categorized as labels and messages, whereas support information is categorized as online and printed information. Usability specialists have always recognized the importance of information and the usability heuristics they have formulated include the information component. For example, Molich and Nielsen's 10 software usability principles mention all four of the information components that we earlier identified. Table 1.2, which lists the relevant principles, clearly shows how information contributes to software usability.

1.4.2 What are information elements?

Before we move further, let us define the term information element. An *information element* is a unique "package" of information specifically designed for the targeted users, medium, and purpose. For example, *OS/2 Warp User's Guide* is an information element. Messages may actually come in one or more information elements. For example, the following three types of messages are unique information elements: informative messages, warning messages, and action messages. Another example is the various types of online Help you may provide: field-level Help and message Help are two different information elements. See Figure 1.6.

Table 1.2
Molich and Nielsen's Usability Principles With the Corresponding
Information Components

Usability Heuristic	Information Component Indicated
Component 2. Speak the user's language. The text should be expressed clearly in words, phrases, and concepts familiar to the user, rather than in system-oriented terms.	Labels and messages
Component 5. Feedback. The system should always keep users informed about what is going on, through appropriate feedback, within reasonable time.	Messages
Component 8. Good error messages. They should be expressed in plain language (no codes), precisely indicate the problem and constructively suggest a solution.	Messages
Component 10. Help and documentation. Even though it is better if the system can be used without documentation, it may be necessary to provide it along with Help. Any such information should be easy to search, be focused on the user's task, list concrete steps to be carried out, and not be too large.	Online and printed support information

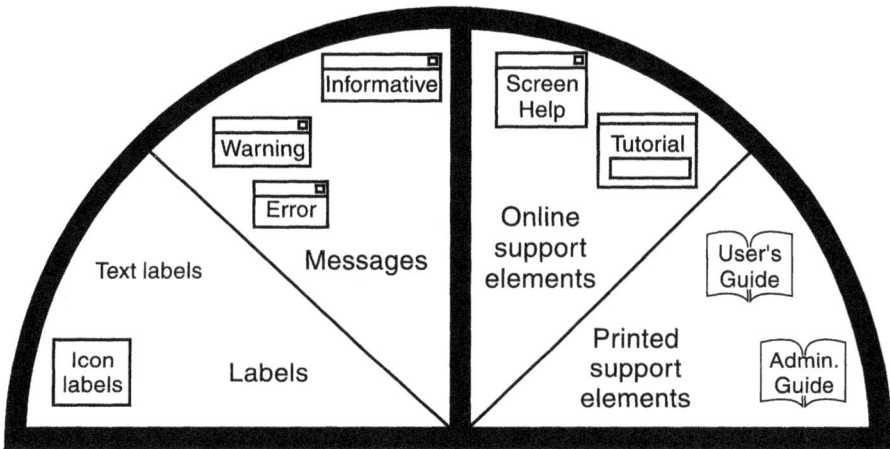

Figure 1.6   Examples of information elements.

1.4.3   Interaction information

Labels and messages are information, and are essential for the completion of user-software interaction. Therefore, we will call them *interaction information*. Horton [5] calls messages and labels an "adjunct form of online documentation" and "preemptive documentation" because information communicated clearly by messages, commands, and menus need not appear in online Help or any other information element. If the quality of interaction information is good, users most likely will have less need for support information.

1.4.3.1   Labels

A *label* is the name that identifies a user object. For example, the option "Edit" in your word processor is a label. Here are some examples of user objects that need labels:

- windows;

- menu options;

- fields;

- commands.

A good label helps users quickly "retrieve" a user object they are looking for, and helps them understand what the object is. Therefore, the label is an important software usability component.

A label can be a word or phrase, a word or phrase plus description, an icon, or a combination of these. A brief description alongside a textual label has often helped users of character-based user interfaces. In graphical user interfaces, icons are increasing used. Though often helpful, you cannot entirely rely on icons to "convey the information." Therefore, either a combination of icon and text is used or the text appears in some form of online Help such as the balloon Help in windowing systems.

Labels, especially those for menu options, usually have keyboard shortcuts associated with them. For example, the abbreviation for File Open could be CTRL + O, where "O" is for Open. See Chapter 9 for a detailed dissucssion on designing labels.

1.4.3.2 Messages

A software program's reaction to a user action can be in terms of presenting the requested results or providing feedback. We are here concerned about feedback, which is an information component. In this book, we will call it a message. A *message* can be provided in different ways. It can be textual, graphical, audible, or a combination. Common types of messages are informative, warning, and action. See Chapter 10 for a detailed discussion on designing effective messages.

1.4.4 Support information

Even though it is preferable if a system is so easy to use that no further help or documentation is needed to supplement the user interface itself, this goal cannot always be met [4].

Online support information and *printed support information* "clarify" the other software usability components based on a user's request or need.

Does a better design of the interaction components mean less need for support information? Yes. Earlier, when personal computers were sold, they came with a big instruction manual on how to assemble the computer. Then the user interface was improved. Computers then came assembled. Now all that users need to do is plug things in. The result is an instruction foldout rather than a bulky manual.

However, the user interface probably cannot be improved to the point of not needing any support information at all. The exceptions may be "walk-up-and-use" systems such as automated teller machines. The "no support information" slogan probably cannot be realized for software with sophisticated features. Users will need support information for a clarification on the other software usability components.

To my employer's World Wide Web site, we wanted to include a guestbook. We came up with a perl script that would allow site visitors to enter and submit their details. The script did not work. In order for it to work, we needed to make some changes in our Web server. We knew it could be done with this server, but never figured out how because we did not have the manuals. The result: we bought another server. The moral of the story: support information is important for users.

1.4.4.1   Online support information

Support information appears on the user's screen. Some online support information elements are as follows:

- README files;
- Online manual;
- Field Help;
- Message Help;
- Online tutorial;
- Cue cards;
- Demos;
- Examples;
- Web-based information elements.

See Chapter 11 for a detailed discussion on designing online information elements.

1.4.4.2   Printed support information

Support information is printed in the form of manuals, cards, or foldouts. Some printed support information elements are as follows:

- Guide;
- Reference;
- Reference card.

See Chapter 12 for a detailed discussion on designing online information elements for the print media.

References

[1]   Galitz, Wilbert O., *Handbook of Screen Format Design*, Second Edition, Wellesley, MA: QED Information Sciences, 1985.

[2]   Reed, Sandy, "Who Defines Usability? You Do!" *PC Computing*, Dec. 1992.

[3] Dayton, Tom, et al., "Skills needed by user-centered design practitioners in real software development environments," Report on the CHI '92 Workshop, *SIGCHI Bulletin*, July 1993, Vol. 25, No. 3.

[4] Nielsen, Jakob, *Usability Engineering*, Cambridge, MA: AP Professional, 1993.

[5] Horton, William, *Designing & Writing Online Documentation: Help Files to Hypertext*, New York, NY: John Wiley & Sons, Inc.

2

What is UCID?

Tᴴɪꜱ ᴄʜᴀᴘᴛᴇʀ introduces *user-centered information design* (UCID) and describes the key features of this approach. You will learn what UCID can do for you and how you can integrate it into your current software engineering process.

2.1   How information is developed, *traditionally*

Everyone appears to complain that information quality is poor. Take a good look at the literature and the currently practiced approaches, and you will realize that there is a bunch of misconceptions and inadequate processes out there. I have described some of them here to show how they do *not* help if your goal is to enhance software usability.

The "documenting the software" approach Those who practice this approach say, "We've written everything about the software." The good thing about this approach is that *all* information is provided. When you document software, you write everything about the software as long as time allows. You see this approach in the large volumes of manuals written entirely from the programming viewpoint. The drawback of this approach is that there is no concern for the users—people who will actually use the software and information. The right approach would be to provide only the information users need and expect.

The writer-centered approach The focus here is on things like grammar, rhetoric, writing style, mechanical style, typography, and activities like word processing, copy editing, printing, and binding. The resulting manuals are strong in these areas. However, such "well-written and produced" manuals are about as useful as advertisements that win awards but don't sell products. People are not buying literature. They are buying software products to make their lives and their customers' lives easier. The right approach then would be a user-centered approach.

The "writing usable manuals" approach Here, information is thought of as a separate entity that deserves its own *independent* set of methods for development, evaluation, and so forth. So, among other things, you have usability tests specifically designed for testing information alone. This is an improved approach, where the usability of individual information elements is improved. However, together, the information elements may not significantly contribute to improved software usability. Moreover, support information elements may not be consistent with the rest of the product. For example, online Help may have its own user interface—different from the product's. The right approach, then, would be one that designs information in the context of software usability.

The form-based approach Writers implement preconceived ideas of what information elements should be delivered along with software. Many books list and describe a *User's Guide* and a *Reference Manual* as the standard information elements to be delivered with any software. A custom software development company's in-house guidelines document recommends a *User-Cum-Operations* manual. The good thing about this approach is that you avoid analysis work and save time. All you have to do is tailor information to fit the preset forms. The problems with this approach are many. Information that users seek may not be there. Or users may seek certain types of information in the wrong information element—what they quickly need on the computer is probably hidden in a 400-page printed manual, resulting in time wasting and user frustration. Or information is not provided in a form that users expect. The right approach is to decide on information elements based on a knowledge of your users, the tasks they perform with the software, characteristics of online and printed media, and the complexity of the user interface.

The print-oriented processes Most traditional processes are meant for "producing" printed manuals. Today, however, some form of online support information (e.g., Help) is almost always expected by users, and therefore delivered by software houses. The right process would then allow for the design of online support information elements and their integration with printed support information elements.

The "information is documentation" thinking Today, the word *information* is preferred to documentation, but is only used to mean online and printed support information. Actually, software usability is more severely affected by two other information components—labels and messages—the design of which requires good technical communication skills.

The "technical writers write only documentation" thinking Software engineers traditionally have written labels and messages. For effectiveness, both of these information components need that critical user viewpoint thinking and writing that good technical writers are capable of.

The "analyze users and tasks" advice This is good advice. Information design should be based on a knowledge of users and tasks. Unfortunately, this advice appears in literature aimed at technical writers. If technical writers only write documentation, such user/task analysis results are more important for usability engineers and user interface designers, who are also better informed in analysis methods. If the role of technical writers is extended to include the design of interaction information (labels and messages), then the right approach would be a process where usability engineers and technical writers collaborate in analyzing users and tasks.

2.2   User-centered information design

The software technical writing profession has evolved from software-centered "documenting the software" to writer-centered "writing well," to document-centered "writing usable manuals," and now to the software usability-driven approach described in this book. See Figure 2.1.

With software usability enhancement as its goal, UCID integrates the design of all the information elements that users require. UCID looks at the role of information from the context of product usability—a holistic view (see Figure 2.2). Therefore, it covers all four of the information components that impact software usability: labels, messages, online support information, and printed support information.

UCID should be practiced in software engineering projects that follow the user-centered approach to designing the product's user interface. Technical writers are the primary contributors to UCID, enjoying an expanded role that includes the design of labels and messages, besides support information. For many activities, the UCID process gets technical writers, usability engineers, and programmers to collaborate toward achieving the goal of improved software usability.

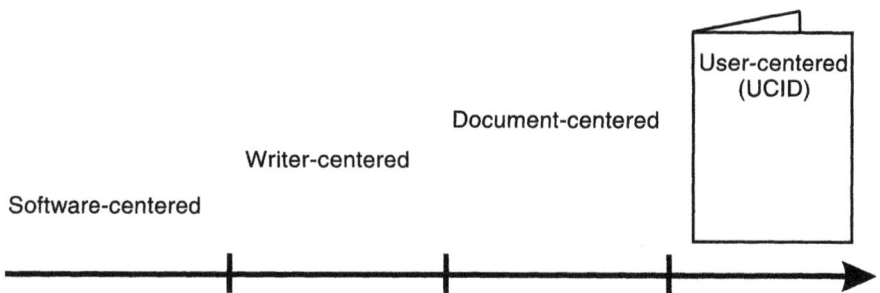

Figure 2.1   Evolution of software technical writing approaches.

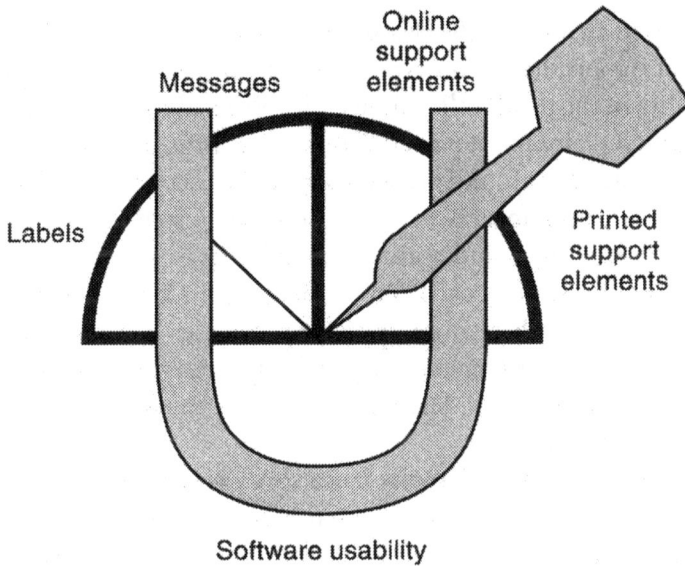

Figure 2.2   User-centered information design.

2.3   Five key principles of UCID

UCID is a result of information design lessons I learned from experience and from literature.

Lessons I learned from experience   A multinational giant outsourced the development of a large accounting software product to a company where I previously worked. The project spanned over three years. One of the requirements was that the product should be high on usability. The user interface had to conform to our client's standards. In addition, three printed manuals and a large online Help system were to be delivered. Interestingly, we did not have in-house usability or user interface design skills, not even professional technical writing skills. Because I had a technical writing background, I was quickly recruited into this project. My responsibilities were clear: train programmers to write manuals, guide them as they write, edit their work, and manage the manual development project. In the beginning, everything went fine.

With help from our client's usability specialists, we lab-tested the usability of the product prototype. Trouble began. The client screamed at the quality of the product's online Help. We had to change the process. Online Help development passed hands to my information development team. Soon, the client found problems with the terminology (labels) used in the product user interface. The client suggested that I review the labels. I reviewed them and realized I should have done it myself right at the beginning of the project. Then we delivered a large number of error and other messages that could appear on the user's screen. The client expressed shock. The project manager thought maybe I could review the messages as well. Actually, I had to rewrite the messages in consultation with the programmers.

I learned two critical lessons from this project experience. Lesson one: good technical writing skills are needed for the effective design of labels and messages. Lesson two: in order to avoid incomplete and inconsistent information, all the information elements should be designed together with improved software usability as the goal.

Lessons I learned from literature Usability professionals have always believed that information elements that software houses deliver impact software usability. In their set of 10 software usability principles [1], Molich and Nielsen have given significance to the quality of information components. Jeffrey Rubin [2] and others have even pointed out the need for the integrated design of various information components.

UCID gets over the limitations found with traditional approaches, some of which we have seen earlier in this chapter. Figure 2.3 shows five principles that embody UCID and distinguish it from other approaches.

2.3.1 Keep software usability the goal—always

One of the goals for the software engineering project team is enhanced software usability. All related activities are aimed at achieving that goal. Being a software usability component, information cannot be seen and designed in isolation from the broader product usability context. The *primary* purpose of every information element you provide is to work along with the other information elements—even with the other software usability components—to meet the goal.

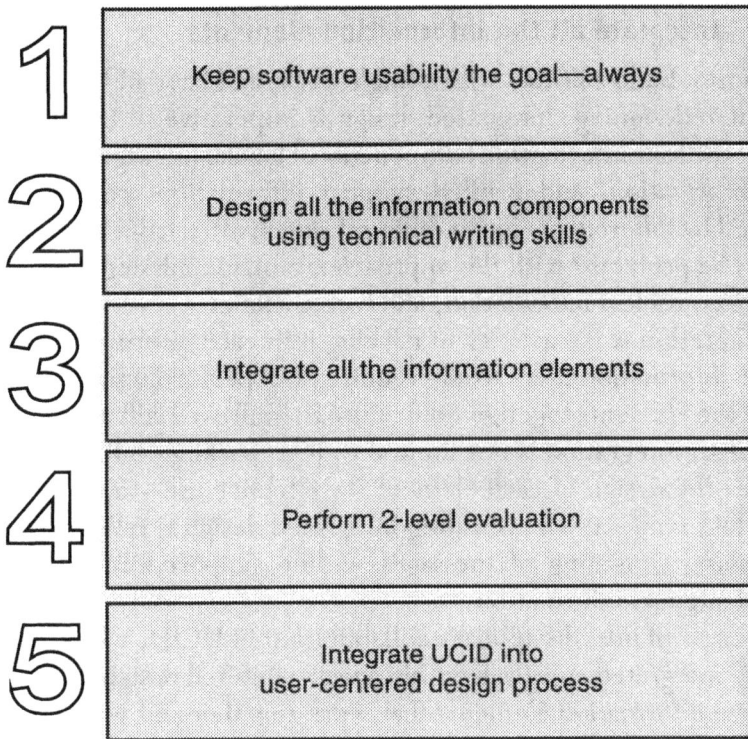

Figure 2.3 Five key principles of UCID.

2.3.2 Design *all* the information components using technical writing skills

In Chapter 1, we identified and categorized all the information components that impact software usability. Besides online and printed support
information, the list includes labels and messages. However, technical
writers typically only design support information.

Good technical writers write support information well because they
understand programming and application domain details, think from
the user's viewpoint, and have good written communication skills. Interestingly, the same skills set is required to better design interaction
information—that is, labels and messages. UCID therefore expands the
technical writer's role and brings those skills to the process of interaction
information design as well.

2.3.3   Integrate all the information elements

You cannot build a bridge with a single deck, and later add another deck without redesigning. Integrated design is imperative to UCID as well. Each of the four information components of labels, messages, online support information, and printed support information are traditionally designed by different groups at different times, often from different locations. The problems with this approach are many: missing information, repetition, useless information, and inconsistency.

Integration is the activity of packing into each information element just the information users would require or expect from that element so that all the elements together contribute to improved software usability. Of course, integration is not limited to just "packing" information, but includes the design of each element for effective use—considering the users' task context. The resulting integrated design is called *information architecture*, consisting of messages, online support information, and printed support information.

Because of interdisciplinary collaboration in UCID, when you practice this integrated approach within a user-centered design process, you will have information elements that work together and with the rest of the software usability components. Users will perceive the software system as one single usable system.

2.3.4   Perform two-level evaluation

Follow an evaluation approach that will improve individual information elements so that they will, together and with the interaction components, maximize software usability.

This approach requires two levels of evaluation. The first level, called integration evaluation, involves the testing of the entire software system. All the software usability components are—as one piece—tested for usability by test participants, who are actual users or representative users.

The second level, called *critical design goal* (CDG) evaluation, involves the tests, reviews, and editing of each information element. Here, evaluation is primarily to determine to what extent you have achieved the CDGs defined earlier in the project for each information element. The CDG for an information element is a set of information design goals essential for the success of that information element in contributing to improved software usability. CDGs for labels, messages, and online Help are described in Chapters 9, 10, and 11, respectively.

2.3.5 Integrate UCID into the user-centered design process

The UCID process fits into, and should be applied in, the user-centered design process. It shares some activities with other traditional processes, but is still unique in many ways, as described in Chapter 3.

2.4 Getting UCID into your organization

If your organization gives importance to good user interface design and you want to maximize software usability by improving information design, you are probably ready to seize UCID. The following text describes what UCID can and cannot do for you. If you believe you are ready, learning and applying it is simple, as shown later in this section.

2.4.1 Know UCID advantages

UCID is the answer to the problems listed in the first section of this chapter. Here is a summary of its advantages, both for users and practitioners.

Advantages for users include:

- Helps maximize the ease of use of, productivity of, and overall satisfaction with the software.

- Avoids *unnecessary* repetition of information, elimination of critical information, delivery of useless information, and inconsistencies among the software usability components.

- Wins the preference of users because they become involved in, and can contribute to, the design of information elements.

Advantages for practitioners include:

- Puts you on the right track toward maximizing software usability.

- Improves company's bottom-line. Information meeting users' needs and expectations means improved software usability and the resulting positive impact on the company's revenues.

- Gives possible competitive advantage to organizations that take it up first.

- Reduces costs by resolving design flaws at an early stage and by avoiding costly corrections toward the end. Here, a design flaw can be a wrong set of information elements or an information element that is badly organized.

- Plays an important role in establishing the identity of technical writers as designers of product usability.

- Improves project communication. UCID requires that you record details such as the design rationale. These details help writers who join the team at a later stage.

2.4.2   Know UCID limitations

UCID will not make sense in a software engineering process that does not have a user-centered approach to designing the user interface. Moreover, UCID is only as effective as the expertise and availability of technical writers, usability engineers, and other required resources. For example, writers who poorly write manuals are likely to poorly write messages too. It sure is not a good placement for a recent graduate.

The UCID approach is simple. However, it is not simple enough to be used in very small projects of, say, under four months elapsed time. Moreover, activities such as the definition of CDGs and the two-level evaluation take relatively more effort.

2.4.3   Steps to implementing UCID

Here are the steps for the successful implementation of UCID in your organization:

1. *Get support*. Get formal management commitment. If you already have user-centered design in your organization, it becomes easier to get management to extend support for UCID. Clarify to management UCID's advantages, both for users and practitioners.

You will also need the support of the usability and programming teams. Your objective should be to help them realize the need to rethink the ways of improving software usability. In particular, you should help them understand that information impacts software usability.

2. *Expand roles.* The usability manager/engineer should integrate the UCID process into the user-centered design process in consultation with the software project manager. He or she should also have the new responsibility of leading the technical writers. (This leadership is at the project level and the technical writing group will continue under its current technical writing manager, thus retaining the advantages of specialization.) Usability engineers experienced in analysis should also initially handhold technical writers to collaborate in analysis and evaluation activities.

Expand the role of technical writers in your organization to include designing labels and collaborating to write messages, conduct analysis, and set goals. Get them oriented to usability engineering principles in-house by your usability engineers.

3. *Prepare to collaborate.* Set up a working relationship among technical writing, programming, and usability teams. Let everyone know what each is expected to do.

4. *Integrate UCID into user-centered design.* Let the usability manager/engineer, who leads the user-centered design effort, integrate UCID in consultation with the software project manager. Also, update your corporate process documents to reflect this integration. Now you are ready to try your pilot UCID project.

References

[1] Nielsen, Jakob, *Usability Engineering*, Cambridge, MA: AP Professional, 1993.

[2] Rubin, Jeffery, *Handbook of Usability Testing: How to Plan, Design, and Conduct Effective Tests,* New York, NY: John Wiley & Sons, Inc., 1994.

3

The UCID Process

THIS CHAPTER starts with a brief description of traditional processes
for information development. Since its goal is to enhance software
usability, the UCID process only makes sense in a project that
follows the user-centered approach to designing the user interface.
Therefore, this chapter also discusses user-centered design.

This chapter also describes the key features of the UCID process,
showing how it overcomes the limitations of current processes. In
conclusion, we go into specific UCID activities.

3.1 Limitations of traditional processes

Authors and software organizations have defined and applied processes
for information development. Some of these processes, especially those
defined as corporate processes, are excellent in getting information development done efficiently. They incorporate tight project management

controls so that things get done on time and within budget. However, from a user viewpoint, they are inadequate because:

- Current processes are oriented to the production of printed support elements. In fact, technical writing books have been written to help *produce* manuals. Today, this print orientation is inadequate, as users expect a lot of information online.

- Current processes do not integrate the design of online and printed support elements. Even in large software organizations, separate teams develop support elements for the two media independent of each other, and often from different geographic locations. With such lack of integration, organizations can hardly meets users' information needs and expectations.

- Current information development processes do not include labels and messages, which would also greatly benefit from the use of technical writing skills.

The UCID process ensures the integrated design of all the information elements.

3.2 User-centered design

The UCID process is followed in user-centered design projects. It is, therefore, important to understand the concepts and process of user-centered design.

User-centered design (Figure 3.1) is a set of activities that enhance software usability through implementing the following key concepts:

- *Focus early on users and tasks.* The aim is to understand users' cognitive, behavioral, and attitudinal characteristics, the tasks users perform, how, and in what kind of environment. This understanding is gained via direct contact with actual or representative users and via indirect sources.

- *First design the user interface.* User interface design is separated from, and precedes, internal design. The software system's internal

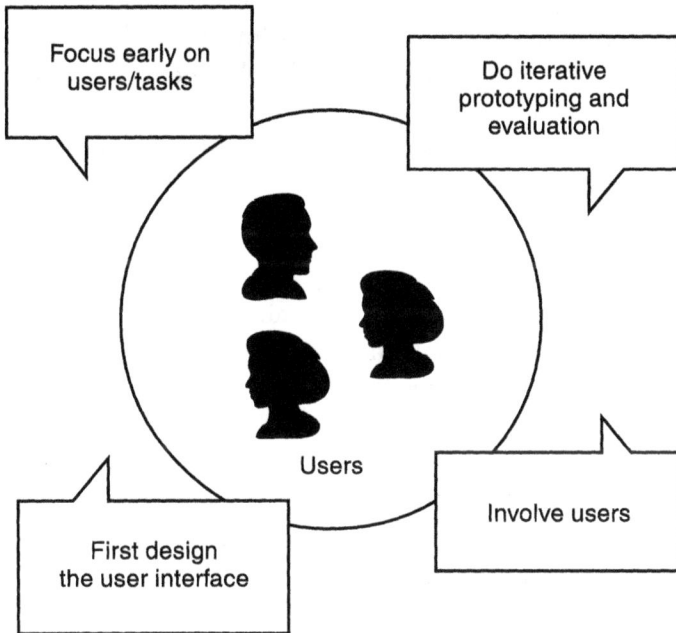

Figure 3.1 User-centered design.

implementation is structured so that the user interface can be changed without changing the internals.

■ *Involve users.* The extent of user participation varies. An evaluation method, such as walkthrough, brings users into the evaluation process while the participatory design method allows users to be part of the design team, involving them in design and evaluation.

■ *Insist on iterative prototyping and evaluation.* The design of the user interface is evolved via iterative prototyping. Based on measurements and observations made during user testing, the prototype is redesigned. The prototype-test-prototype cycle continues until software usability goals defined earlier in the process are met.

By applying these concepts, user-centered design ensures that software systems are closely consistent with users' needs, expectations, and characteristics.

User-centered design (Table 3.1) fits into the software engineering process that consists of analysis, design, development, and testing phases. This process is often led by a usability manager or engineer who has management skills as well as knowledge and experience in human-computer interaction techniques.

3.2.1 Analysis

The analysis phase comprises activities aimed at increasing the design team's knowledge of the people they will be designing for, the tasks these people will perform with the software, and the environment where the software will be used. The specific activities are described below.

3.2.1.1 Analyze users and tasks

Activities are carried out to learn about the tasks to be performed with the software, and about the users who will perform the tasks. Methods are

Table 3.1

User-Centered Design Process

Software Engineering	User-Centered Design
Analysis	**Analysis**
	Analyze users and tasks
	Set software usability goals
	Plan for software usability
Design	**Design/prototyping**
	Design/prototype user interface
Development	**Implementation**
	Code user interface
Testing	**Evaluation**
	Perform usability inspections
	Perform usability testing

many, ranging from simple discussions with a few prospective users to performing field research, called contextual inquiry, where users are observed and interviewed in their actual work environment.

The result of user analysis is a *user profile* that describes the characteristics of the users. And the result of task analysis is a *task description* that explains what tasks the new software should support and how users should perform those tasks in terms of inputs (user actions) and outputs (reports, etc.).

3.2.1.2   Set software usability goals

Measurable (and other) goals are set to guide the design and evaluation efforts. An example of a software usability goal is that users should be able to install the product in under 45 minutes. Another example could be that the number of errors made by users while performing a task should be less than three. Software usability goals can be based on the performance of similar or competing products or on the organization's judgment of usability needs.

3.2.1.3   Plan for software usability

Planning answers the question, "How do we evaluate the software's usability?" It lists the evaluation methods to be used and describes how and when these methods will be applied.

3.2.2   Design/prototyping

A prototype of the new user interface is designed based on the knowledge gained from analysis, and on the software usability goals. This is a critical activity that decides the "look and feel" of the software. The prototype can be a paper prototype showing screens and navigation or a working model developed using software tools. It is improved through a prototype-evaluate-prototype cycle until the design meets the defined software usability goals (or satisfies users' needs and expectations.)

To design the user interface, one or more of the many design techniques is followed. One design approach is called participatory design, where users are part of the design team and are involved in the design effort, participating in meetings and evaluations. Often, two or more designers simultaneously work to come up with different design ideas.

This helps evaluate multiple designs and pick the one that users are most comfortable with.

Design rationale, the reason behind the design idea, is documented. It helps when you want to change the design.

3.2.3   Implementation

The result of prototype-evaluate-prototype iteration is an approved prototype. The approved prototype is translated into a full working user interface through coding or through the use of software tools. Some user-centered design process models include a phase preceding implementation. This phase may involve activities such as writing detailed specifications or creating state diagrams.

3.2.4   Evaluation

The software engineering team plans for usability evaluation during the analysis phase. Evaluation is an "umbrella" activity that happens throughout the user-centered design process. It is done through two broad methods, usability inspections and usability testing.

3.2.4.1   Perform usability inspections

Robert Mack [1] defines inspection as a set of methods based on having evaluators inspect or examine usability-related aspects of a user interface. The methods are all inexpensive and simple to use at various stages of user-centered design. Usability inspectors can be usability specialists, software engineers, or users. On completing their inspection, these evaluators list all the usability problems and also provide their recommendations for removing the problems. Some of the commonly used inspection methods are described in the following paragraphs.

Heuristic evaluation   Usability specialists evaluate to see if software usability components—user objects, user actions, and messages—conform to established usability principles (or heuristics). An example of a usability heuristic is "Provide adequate feedback." Usability specialists use this heuristic to check if the software provides adequate messages, progress indicators, and so forth.

Walkthrough In what is called pluralistic walkthrough, usability specialists, programmers, and users meet to walk step by step through a task scenario, discussing usability issues. In another technique, called cognitive walkthrough, the creator of one aspect of a design presents it to peers, who evaluate the proposed design in the context of tasks users will perform. This technique focuses on evaluating the product's ease of learning.

Standards inspection Standards are things like IBM's *Common User Access* (CUA) and Microsoft's Windows standards. Experts in a particular standard inspect the user interface for conformance to that standard.

3.2.4.2 Perform usability testing

The prototype or product is tested against the software usability goals defined in the analysis phase. Actual or representative users perform tasks with the user interface. The test is observed and possibly recorded, then played back and reviewed. Users' problem areas are identified, and solutions are recommended and implemented. Perhaps the most common testing technique is a laboratory test. Jeffrey Rubin [2] describes six stages for lab testing: develop test plan, select and acquire test participants, prepare test materials, set up test environment, conduct test, and debrief test participants.

3.3 The UCID process: key features

The UCID process shares some characteristics and activities with traditional information development processes. The process is simple. The software organization can quickly prepare, and the process can be easily applied. Here are the key features of the UCID process:

- Should form part of the user-centered design process and be managed by a usability manager/engineer.

- Should cover all the four information components that impact software usability. This includes labels and messages, because the skills required for better design of interaction information are the same technical writing skills required for writing support information.

- Should *integrate* the design of all the information elements. Should ensure integration through:

    Collaboration: Coperformance of analysis, goal setting, and other activities by usability engineers and technical writers. Collaboration also helps technical writers learn analysis techniques and design better labels, messages, and other information as they stay focused on software usability.

    Integration evaluation: Simultaneous evaluation of all the software usability components to check if information design meets users' requirements while they perform tasks with the software.

- Should ensure that design is good from the users' viewpoint. A knowledge of users' requirements and characteristics is used in design. The UCID process also continually seeks user input during the prototype-evaluate-prototype cycle. Moreover, it requires approved specifications and prototypes before you can actually start writing.

- Should promote specialization. Labels and messages are designed by technical writers, who are specialists in user-oriented thinking and writing. Moreover, the application of the critical design goals technique promotes the quality of individual information elements.

3.4 The UCID activities

The UCID process (Table 3.2) fits into the user-centered design process described earlier in this chapter. It consists of four phases in sequence: analysis, *high-level design* (HLD), *low-level design* (LLD), and development. Evaluation is an "umbrella" activity appearing in all these four phases.

Each phase, including evaluation, has a set of activities. Each is completed only when evaluation criteria are met, which is usually through iteration. There are thirteen activities in UCID, and they all must be performed. The thirteen activities produce seven deliverables and one project document, called the UCID plan. Though all the UCID phases and

Table 3.2
The UCID Process

User-Centered Design Phases	UCID Phases/Activities	Deliverables	Project Documents
Analysis	**Analysis**		
	1. Coanalyze users and tasks		Usability plan
	2. Cospecify software usability goals		Usability plan
	3. Develop UCID plan		UCID plan
Design/ Prototyping	**High-level design**		
	1. Design labels	Labels	UCID plan
	2. Design information architecture	Information architecture	UCID plan
	3. Plan for information quality		UCID plan
	Low-level design		
	1. Design the specifications	Specifications	UCID plan
	2. Design prototypes	Prototypes	UCID plan
Implementation	**Development**		
	1. Develop messages	Text and code of messages	
	2. Develop and implement online support information elements	Text and code of online support elements	
	3. Develop printed support information elements	Drafts of printed support elements	
Evaluation	**Evaluation**		
	1. Perform integration evaluation		UCID plan
	2. Perform CDG evaluation		UCID plan

activities are mandatory, you have the flexibility to start and end activities as required by your organization's user-centered design process.

3.4.1   Analysis

Analysis is a set of groundwork activities that provide key details, based on which the software (including its information components) is designed.

The UCID analysis phase is the same as the analysis phase in user-centered design. The only difference is that technical writers work closely with usability engineers, provide inputs to all the activities, and prepare an UCID plan. This collaboration helps in the total integration of all the software usability components.

Probably the only software project documents available at the start of this phase are the product proposal and initial drafts of requirements specifications and project plan. At the end of the phase, you will have the initial drafts of the usability plan and UCID plan.

The analysis activities are described next.

3.4.1.1   Coanalyze users and tasks

Together with usability engineers, technical writers should get into direct contact with prospective users to learn about their characteristics and their tasks. Technical writers should also record the user profiles and task descriptions in the usability plan. *User profiles* describe the characteristics of the users, whereas *task descriptions* list what tasks the new software should support and how users should perform those tasks in terms of inputs (user actions) and outputs (reports, etc.). See Chapter 5 for a detailed discussion on analysis.

3.4.1.2   Cospecify software usability goals

Technical writers should assist usability engineers in setting software usability goals. They record the goals and other usability requirements in the usability plan. See Chapter 6 for a discussion on defining software usability and information quality goals.

3.4.1.3 Develop UCID plan

Record all details related to information design in the UCID plan. This document, developed by technical writers, should evolve throughout the software engineering project. See Chapter 4 for a discussion on the UCID plan.

3.4.2 High-level design

HLD is perhaps the most critical UCID phase because it is during this phase you decide what you will provide in what form, and in which media. Any flaws introduced here will be hard to correct later.

Spanning the design/prototyping phase of user-centered design, the UCID HLD phase starts when you have a user interface prototype or a draft of functional specifications. Of course, analysis should be complete and software usability goals should be available.

The HLD activities are described next.

3.4.2.1 Design labels

Prepare a glossary of all the terms, most of which will appear as labels for various user objects. Evaluate and iterate the term definition activity. Document the resulting approved glossary in the information architecture section of your UCID plan. The glossary should be used and updated as required. For more details, see Chapter 9. When you have finished with the glossary, you will probably only deal with the other three information components: messages, online support elements, and printed support elements.

3.4.2.2 Design information architecture

Identify all the information elements that users need and expect based on the following:

- Users' information requirements, which you will identify using the results of user/task analysis and task (or user interface) complexity;

- Your knowledge of the characteristics of online and print media;

- Project-specific constraints and requirements.

The resulting information architecture covers messages, online support elements, and printed support elements. (It will also contain the glossary of labels you defined earlier.) In the information architecture section of the UCID plan, briefly describe the relationship among the information elements. Then, for each information element to be developed, give these details: title, objectives (how it meets user needs, where in the information architecture it fits in), users, high-level content, page/screen count, prerequisites and dependencies, and production details (color, graphics, etc.). With such details, you have the design rationale, which will be very helpful when new technical writers join in later in the project, as well as when you start maintaining information after product release.

The information architecture is iteratively designed and documented in the UCID plan. For more details on information architecture, see Chapter 7.

3.4.2.3   Plan for information quality

The following activities are part of the UCID quality process:

1. Set goals;

2. Design, keeping the goals in mind;

3. Evaluate, to see if goals are met;

4. Improve;

5. Iterate.

Planning for information quality involves setting goals and defining evaluation activities. Information is good quality only when it contributes to improved software usability. Define information quality goals as explained in Chapter 6. Then plan to evaluate how well information meets the goal of improved software usability. Specifically, plan the methods and techniques you will use for evaluating information at two levels. The first level involves evaluating how well all the information elements are integrated. The second level involves evaluating individual information elements against the corresponding critical design goals you defined during the goal-setting process. Record your planning details in the UCID plan.

3.4.3 Low-level design

LLD involves the design of the details of every information element listed in your information architecture. It clarifies the exact "look and feel" of each information element, helping you get the required approvals.

LLD spans the later part of the design/prototyping phase of user-centered design, probably even creeping into the development phase. LLD starts when you have a fairly complete set of functional specifications.

The LLD activities are described next.

3.4.3.1 Design the specifications

Specifications at this stage are for support information elements. Using detailed and descriptive outlines, specify the content and organization of each information element. Each outline should list the topics to be covered in the order in which they will be covered. An important thing to remember here is to focus on the quality of headings, which makes it easy to retrieve the information users seek. Record your specifications in the UCID plan.

Specifications help (1) obtain early agreement on the content, (2) avoid reorganizing the information element late in the project, (3) allow two or more writers to work on the same element, and (4) add a new writer in the middle of the project. See Chapter 8 for a discussion on designing your specifications.

3.4.3.2 Design prototypes

Design multiple prototypes for each information element defined in the information architecture, even for each type of message. Iteratively evaluate and redesign, and pick for each element one prototype that best meets users' requirements. Use the approved prototypes as models for developing the information elements. For example, the approved prototype for a printed guide can be a representative chapter containing all typical types of information and other items you would include, written and formatted the way the rest of the chapters should be. Document the prototypes in the UCID plan.

Prototyping is a low-cost method for getting feedback before you invest time and other resources for, say, developing a large draft of a printed guide. See Chapter 8 for a discussion on designing your prototypes.

3.4.4   Development

Development is the writing and illustrating of various drafts of each information element based on specifications, prototypes (models), and corporate style guides, if any. It also includes storyboarding, coding, and other activities to produce complete and working information elements.

All the support information elements are evolved through various drafts. A three-drafts approach is recommended. UCID is not strict on what constitutes each draft. It just emphasizes iteration appropriate for your project.

The development phase spans the coding and testing phases of the software engineering cycle. Enter this phase only after your specifications and prototypes have been approved via design-evaluate-design iteration. At this stage, you should have these source materials: functional specifications, a system design document, and programming specifications.

3.4.4.1   Develop messages

Programmers best know why a message occurs and how errors, if any, could be corrected. Technical writers, who think from the users' viewpoint, can better communicate the information users need. Therefore, technical writers should work closely with programmers to coauthor messages.

3.4.4.2   Develop and implement online support elements

Use specifications and prototypes to write drafts of all the support elements meant for the online medium. Implementation includes activities such as coding required to produce the final working element. Traditionally, implementation has been done by programmers. Today, with the availability of easy-to-use development tools, writers increasingly "code" the user interface for online support elements.

3.4.4.3   Develop printed support elements

Write drafts of all the support elements meant for the print medium. Write them based on the specifications and prototypes.

3.4.5 Evaluation

Evaluation is not the last UCID phase. Evaluation continually occurs across HLD, LLD, and development phases.

Here are the evaluation-related activities:

- *State evaluation objectives.* Define evaluation objectives based on information quality goals. Evaluation objectives can be the same as the software usability and information quality goals you defined earlier. An example of an evaluation objective could be to compare different online retrieval techniques to know which ones help users find information more quickly.

- *Decide on evaluation methods.* Based on evaluation objectives, you must plan what methods you will use to evaluate each UCID deliverable. Commonly used evaluation methods are lab usability tests, field tests, technical reviews, and editing. The methods you use could be different for each UCID deliverable.

- *Perform evaluation.* Before you can actually perform evaluation, you need to do preparatory work like designing tests, briefing reviewers, and preparing test scenarios. Once evaluation is over, you may want to prepare recommendation reports. In the design-evaluate-design iteration, you stop iterating only when you have met evaluation objectives.

3.4.5.1 Perform integration evaluation

The first level, called integration evaluation, is almost the same as the evaluation in user-centered design—the product (including its information components) is evaluated as a whole. Software usability components are—as one piece—tested for usability by test participants, who are actual or representative users. Integration evaluation is not the first evaluation you perform. Actually, you will first review and edit an information element before you evaluate it as part of the integrated software system.

3.4.5.2 Perform CDG evaluation

The second level, called CDG evaluation, is at the individual element level. You should evaluate each element against its CDGs. For example, your CDGs for labels could be Clear, Distinctive, Short, Consistent, and Pleasing. In CDG evaluation, you should determine—via tests and reviews—whether the labels meet those goals.

References

[1] Nielsen, Jakob, and Robert L. Mack, *Usability Inspection Methods,* New York, NY: John Wiley & Sons, Inc., 1994.

[2] Rubin, Jeffery, *Handbook of Usability Testing: How to Plan, Design, and Conduct Effective Tests,* New York, NY: John Wiley & Sons, Inc., 1994.

4

Managing UCID

THERE ARE CERTAIN THINGS that make a UCID project unique.
For example, new approaches are introduced and people are now
empowered with expanded roles. This calls for a focus on certain
management-related activities. This chapter first introduces the key contributors of UCID, and goes on to describe the following management
activities:

- Managing the collaboration;
- Managing change;
- Preparing the usability plan;
- Preparing the UCID plan;
- Implementing a metrics program.

4.1   The UCID contributors

For UCID, you don't need to bring new experts into the organization. You just have to expand the roles of existing experts and get them to do what they do best—together—to achieve the common goal of improved software usability.

This section describes the responsibilities and any UCID-specific training requirements of people who form the design group—that is, the technical writer, usability engineer, software engineer, and user. It also covers other contributors, such as editors. See Figure 4.1.

4.1.1   Technical writers/managers

Technical writers are the primary contributors to UCID. As a technical writer, you will do all the things you have always done—that is, write *support information* such as online Help and printed manuals. However, as a UCID technical writer, you are now creating a part of the product by expanding yourself into *interaction information* design. You design labels for user objects. You also design messages, another important part of the

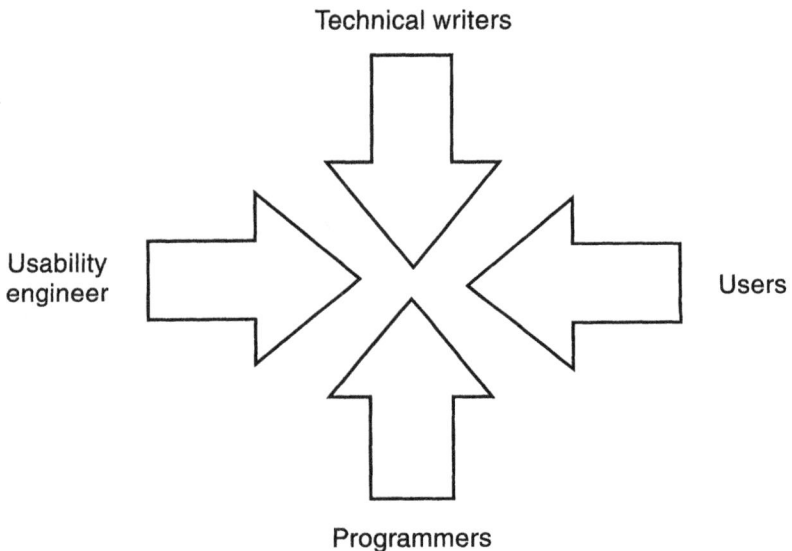

Figure 4.1   The UCID collaborators.

product. This means you will start very early, even before a product prototype exists, and play a role in all phases of the development project.

Besides design and development activities, you will collaborate with usability engineers in performing user/task analysis and in defining goals for software usability and information quality.

Most technical writers will find the expanded role highly motivating and interesting. It makes the best use of your already acquired skills. All you need to do is understand the larger software usability context in which the information components appear. For this, you should try and participate in a training program that covers software usability (or human factors) principles, user-centered design, and task analysis. If you have an in-house usability engineer, training is close at hand. Additional training in collaboration will help. Now being part of the design team, you should develop yourself to think and act like a designer.

The technical writing manager should play a major role in defining the information architecture. He or she should also ensure that technical writers are trained in user-centered design. Note that this management role could also be played by a senior technical writer.

4.1.2   Usability engineers

The usability engineer is variously called usability specialist, human factors specialist, and human factors engineer. Irrespective of the title, the usability engineer is responsible for the usability of the software. Specific activities include analyzing users/tasks, setting software usability goals, conducting training programs in user-centered design, designing user interfaces, and planning for and conducting usability evaluation.

Good usability engineers know the capabilities and limitations of people. They are knowledgeable about user interface design principles, user interface standards, experimental design, and the possibilities and limitations of technology. They also have experience designing user interfaces.

In UCID, the usability engineers coordinate the activities relating to all four components of software usability. Technical writers get to collaborate with them in activities such as setting software usability goals. Some usability engineers have previously had a career in technical writing. For those who do not have this background, it is a good idea to offer them orientation in information design.

4.1.3   Software engineers

One of two categories of software engineers will actually develop the user interface using software tools or traditional programming. The first category consists of people who have got themselves positioned as user interface designers. They are likely to be standards-driven people who are also familiar with user interface design principles. They are typically programmers who have, by virtue of their earlier exposures, become experts in a user interface standard such as Microsoft's Windows standards. The second category consists of "traditional" programmers.

In UCID projects, technical writers should work with software engineers to design labels and messages.

4.1.4   Users

Users may be the people for whom the software is designed. Or they may be representative of people who will use the software. In user-centered design projects, users contribute to both design and evaluation of the software. Some organizations have users as part of the design team, whereas others have them contribute through reviews and tests. Their response in terms of observed difficulty in using the software, or a suggestion for improvement should be considered seriously. In the absence of project-specific constraints, design problems encountered by users should almost always be addressed before the product is released.

4.1.5   Other contributors

Increasingly, graphic designers are becoming a part of the design group. Their role is largely the visual design of screen layouts and icons. They rarely contribute to overall interaction design.

Technical editing and copyediting skills are required to successfully contribute to improving the quality of information. Many editors have both these skills. They need to be familiar with labels and messages. Besides editing, they may also serve other roles, including helping to plan and supervise graphics/print production.

Marketing specialists are important reviewers. They should inspect all the information elements for compliance to the software house's policies. They can also look at the whole product from a user perspective.

The software project manager now has the additional responsibility of understanding and planning for all the UCID activities, ensuring the availability of the required resources, effectively managing the collaboration, and ensuring that everyone in the team understands the expanded role of technical writers. For this, the project manager should first be oriented to UCID and its requirements.

4.2 Managing collaboration

The software development project team traditionally consisted of software engineers (analysts, designers, and programmers), technical writers, and a project manager. The user-centered design approach in recent years has brought usability engineers, users, and even graphic designers into the design team. In UCID projects, technical writers also join the design team.

User-centered design has two characteristics. First, it is performed by a group rather than by isolated individuals. Second, this group is an interdisciplinary team. While collaboration itself can make it challenging to manage a project, interdisciplinary collaboration makes it even more challenging.

The programming and technical writing groups, for example, usually find it hard to understand each other's way of thinking. Software engineers value the program and how it works, whereas technical writers value information and how it communicates. Further, people in different disciplines often come from different departments. Software engineers and project managers are considered the development group. Technical writers may report to a technical writing or communications group. Usability engineers and graphic designers may belong to yet another group. And users—who provide design and evaluation inputs—may be from outside the organization. Such differences give the design group a wide diversity of perspectives, motivation, methods, and priorities.

Collaboration of this interdisciplinary design team is overall managed by the software project manager. To help the manager, we now have the discipline of *computer supported cooperative work* (CSCW). CSCW looks at the way people interact and collaborate with each other and comes up with guidelines for developing technology to assist in collaboration. The technology is called groupware—computer-based tools that advance group work.

4.2.1 Collaboration in UCID

Traditionally, the technical writer collaborated with software engineers. Before writing a manual, the technical writer interviewed software engineers who are subject matter experts to get "raw material" for the manual. A draft written by the technical writer was given to the software engineer for review to ensure technical accuracy. Then, the agreed changes were implemented. Besides collaborating with software engineers, the technical writer also interacted with graphics and print production specialists.

Now part of the design team, the UCID technical writer collaborates with usability and software engineers even more closely than before. What follows is a brief description of the various roles (or people) with which the UCID technical writer collaborates.

Writer-usability engineer Technical writers collaborate with usability engineers in analyzing users/tasks, setting software usability goals, designing the information architecture, planning for quality, and conducting tests. This collaboration helps writers focus on software usability rather than only on the quality of isolated information elements.

Technical writers are more easily accepted by usability engineers than by traditional programmers. This is probably because many usability engineers have themselves had a technical writing career.

Writer-software engineer Technical writers work with software engineers to design labels and messages.

As far as messages are concerned, the software engineer best knows why a message should appear, when it should, and what the user should do in response. The writer is better trained to see from the users' viewpoint what information users will need. Moreover, the writer can present the information in the best possible way. Obviously, a writer-engineer collaboration alone can be highly beneficial for users.

On a large project, my information design team wrote only support information elements. Our multinational customer expressed shock at the messages that went with an intermediate shipment of the product. We changed our message development process for the next shipment. Accordingly, I spent a couple of hours every day with the engineers who coded the messages. The engineer would re-create the message situation, explain it to me, and I would write it out on the computer. The engineer would then read it, say "That's fine" or tell me "No, actually it is this way…" We would then rewrite until both of us were satisfied. You need a level of collaboration similar to this if you want to retain customers the way we did.

Writer-writer This sounds simple because the collaborators are from the same discipline. But there is a different problem here. Because many technical writers share in the writing of various information elements, there can exist a lot of inconsistency. Collaborative writing needs good editors and more editing effort to eliminate inconsistencies and so forth. You should plan for this extra effort.

Writer-editor Farkas and Poltrock [1] report that "In Duffy's survey of 28 expert editors, the ability to establish a collaborative relationship with the writer ranks as number 6 in a list of the 39 most important editorial skills—more important than the ability to find and correct errors of grammar, syntax, and punctuation." The editor should never let personalities interfere with objectivity and the writer should always keep an open mind to the editor's points.

   Successful UCID demands interdisciplinary talents. How can the manager handle this diversity? Keys to the success of such collaboration are acceptance and awareness. In many circumstances, a charismatic leader can generate the required cooperative effort.

4.2.1.1 Acceptance

Each discipline should accept the other for the skills brought to the design group. Technical writers need to be accepted for the unique and important skills they bring to the UCID team. Most usability engineers accept technical writers easily. Many software engineers appreciate them too, but some don't—they may be haughty, and they may have the same attitude towards usability engineers. Traditional programmers may not know about the contribution of these specialists to the organization.

Some may feel that usability design activities are nontechnical and therefore worthless. People with such an attitude can be obstacles to successful user-centered design.

What should the manager do about this? Here are a few suggestions:

- Get software engineers oriented to the principles and practice of good design, with a focus on the interdisciplinary approach. Also, educate them about the work technical writers do, their special skills, and their role in improving overall product usability. Show how their expanded role contributes to meeting user satisfaction.

- For disciplines to work well together, get each designer to learn to appreciate one another's discipline. Encourage people to learn about and value other disciplines. For this, learn a bit about the various disciplines yourself.

- Make it clear that software usability depends also on the contributions expected of programming professionals: the information they provide, the knowledge they share, and their review inputs. Make it clear what each individual is required to do. Document these responsibilities as commitments, with a time frame.

- Continually promote software usability in your organization. Write about its advantages in your company newsletter or intranet, or whichever gets noticed and read. Frequently organize seminars and workshops. Get technical writers as well as usability engineers to cofacilitate these programs.

4.2.1.2   Awareness

Individuals should know what he or she is supposed to contribute and what others in the team do. Where individuals work *closely* together, they typically communicate about their goals, share knowledge, clarify questions, and plan to achieve the goals. Where there is such close collaboration, the manager should set up mandatory communication procedures. In *other* cases, there should still be a mechanism for people to know roughly what others are doing.

4.2.2 Tools for collaboration

Technical writing is often done by a group of people working together. Depending on the number and size of information elements to be developed, and the schedule, each writer may contribute an information element or a section of it. Today, we have computer-based tools, commonly called groupware, to support group writing activities. These include interactive applications for writing, drawing, and e-mailing. Writers can easily collaborate on an information element and even get "online" comments from editors and other reviewers. The computer keeps track of the changes made to information and controls who can change which sections.

Based on the requirements of individuals in the team, the manager should decide on the software tools that will be used for effective collaboration. Here are some of the features you should look for in your groupware application [2]:

- Record and display the identities of the collaborators;

- Support annotations so that editors and other reviewers can record their comments and suggestions;

- Provide the ability to undo changes made by a collaborator;

- Ensure that all users access the same version of the shared document;

- Support the different roles individuals may play, such as specifying content, prototyping, writing, and editing;

- Provide support for multiple document sections (e.g., chapters), yet maintain access to the entire document;

- Provide access to the document concurrently or sequentially;

- Provide feedback to increase the collaborator's awareness of concurrent actions by other collaborators;

- Prevent conflicting changes when two or more people edit a particular part of text at the same time;

- Record who wrote or changed a certain section, what changes were made, and when they were made.

4.3   Managing change

Changes to information elements may be required throughout the UCID process and beyond. They could be triggered either by changes to the software or by information evaluation. Numerous changes to software are typical in real-world projects. New features, bug fixes, and so forth keep coming in. If the changed software is to be usable, the information elements must agree. The iterative nature of user-centered design encourages changes to the "user interface." In addition, you have the changes resulting from the iterative evaluation of information elements themselves.

Changes to information architecture, specifications, prototypes, or drafts can and should be made whenever errors are discovered or suggestions for improvement are received. After shipping, change requests will come from customers. A change anywhere can affect other parts of the integrated whole. Such continuing changes can rapidly lead to chaos—unless managed effectively. That is why we need a change management procedure.

The change management procedure contributes to product quality by formalizing the change request, evaluating it, and controlling its impact. Change management is an "umbrella" activity used throughout the software engineering process. You can combine human tasks and software tools to help control change.

Here is the change management process:

1.   Submit a request for change;

2.   Do casual analysis to document the source or reason for change;

3.   Find solutions to implement;

4.   Evaluate the impact of the solutions on all the software usability components, and on the quality, cost, and schedule;

5.   Select the best solution and make change, if approved;

6.   Correct all the error-producing steps or methods in the software engineering process.

You probably have a change management process already in place. If not, the project manager should ensure that such a process is set up and used.

A related useful process that some organizations follow is the error prevention process. This involves describing the errors found and analyzing the causes, categorizing the errors, and creating suggested actions to prevent them from reoccurring. You can prevent a whole category of errors from reoccurring. Example categories are "Lack of education" and "Lack of communication." For instance, if a support element has incorrect information because the writer did not understand how a parameter worked, the cause of the error is "Lack of education." To better define the category of errors, you should also have subcategories such as "Technical education on product" required in this case.

4.4   Preparing the usability plan

The software usability plan is the framework for designing and evaluating the software to achieve usability. This plan is an evolving document that includes the following items:

- User profiles;

- Task descriptions;

- Software usability goals;

- Features the software requires to be usable;

- Standards and guidelines to be used;

- Plan for information quality.

The usability plan is required for all software projects that follow user-centered design. During the UCID analysis phase, the usability manager/engineer prepares the usability plan in consultation with the technical writer/manager, project manager, and marketing specialists. Have the plan reviewed by the quality assurance manager as well.

4.5   Preparing the UCID plan

The UCID plan consists of the information plan and the project plan. The information plan details what you want to create, whereas the project plan details how you will do it.

The UCID planning process ensures that information elements are not arbitrarily based on what individuals in the project or organization think, but rather based on design iteration and evaluation inputs, especially those from users. The UCID plan clearly shows that a concrete process is being followed to maximize software usability via information design. The plan is useful for the entire project team. It focuses the team's efforts on software usability and helps incoming people come up to speed quickly. Finally, the plan demonstrates the quality of thinking that technical writers bring to the UCID effort.

The UCID plan is required even for the smallest project. Prepare an initial plan during the analysis phase then have it reviewed by the project manager, marketing specialist, usability manager/engineer, and quality assurance manager. Continually improve it based on better understanding of users' information requirements.

4.5.1   The information plan

The information plan section of the UCID plan lists, describes, and justifies the information elements you intend to design. Different sections of the plan are end products of various UCID activities such as goal setting and prototyping. However, if these sections are too large, you might want to make them separate documents.

The information plan should contain the following details:

- Brief introduction to project;

- Brief introduction to product;

- Summary of user profiles (from usability plan, only if required);

- Summary of task descriptions (from usability plan, only if required);

- Software usability goals (repeat from usability plan);

- Plan for information quality;

■ Information quality goals;

■ Information architecture (with design rationale);

■ Glossary of labels;

■ Specifications;

■ Prototypes;

■ Guidelines and standards to be used;

■ Graphics/production/packaging approaches to be used.

4.5.2 The project plan

The project plan section of the UCID plan explains how the information plan should be implemented. Careful planning avoids impossible deadlines and poor-quality information. It promotes a smooth flow of activities, keeping the project on track and schedule.

The project plan should contain the following details:

■ Software project schedule showing key milestones (if required);

■ List of all UCID activities (with brief description if required);

■ UCID team organization, showing collaboration with other teams;

■ Estimates;

■ Dependencies and assumptions;

■ Schedule;

■ Resource allocation;

■ Risk plan.

UCID activities   List and summarize all the UCID activities. Remember to include these activities, too: rework, translation/localization, illustration, and printing.

Team organization   Show team organization. Mention roles and responsibilities of team members and other contributors such as reviewers. Clearly show collaboration. Explain how the team will be managed and how communication will happen among writers, programmers, users, and management.

Estimates   For each activity, estimate the effort, resources, and cost. The fundamental thing to know is "page" count. How would you estimate online information? To make calculations easy, you might want to convert screen count to page count. In one project, we found that two screens of online information were equivalent to one page of printed information. Therefore, the page count for our 100-screen online Help text was listed as 50 pages.

Estimate hours per page for each information element deliverable. See Table 4.1 for sample estimates, which you can use as a guideline. Various factors can impact the estimates for writing depending on the type of information. For example, procedural and conceptual information take more time compared to well-structured reference information. Estimates also depend on the writer's own experience. Some writers need more research and more time with subject matter experts than others. Estimates for editing will depend on the quality of the draft or the skills of the writers.

To get the effort estimate, multiply hours per page by page count. Add appropriate effort for UCID orientation, any new tools to be learned, experience of writers, and so forth.

Dependencies and assumptions   State risks. Mention skills and other resource requirements. State any other dependencies and assumptions.

Schedule   Schedule all the activities based on the effort estimates, available resources, and target date. Have milestones with deliverables and

Table 4.1
Sample Estimates (Adapted from [3])

User's guide	5 hr/page
Reference manual	4 hr/page
Context-sensitive Help	4 hr/topic of Help information
Computer-based training	60 hr/finished hour of training

end dates. You should also have milestones for each draft of every support element. Allow for contingencies. Be prepared to adjust the schedule as the work proceeds and the product changes. Arrange it to meet the overall product deadline.

To keep track, use methods such as time sheets and progress reports. In these documents, record the work done, problems (with reasons), plans, and requirements.

Resource allocations   Allocate appropriate people for various UCID activities. Mention all the contributors, including reviewers. Allocate computer and other resources for each activity, including software tools—both existing and to be procured.

Risk plan   What if the lead writer resigns in the middle of the project? This is a risk item. Identify all the risks for the project. Risks can be related to people, tools, productivity, quality, and schedule. In your risk plan, describe the risk, its impact, and an action plan.

Here is an example of an action plan for the above risk item. You could have one person other than the allocated lead writer "keep in touch" with the UCID project. This person knows the activities being carried out, process, status, and so forth. He or she takes over if the lead writer suddenly resigns.

4.6   Implementing a metrics program

You successfully got management approval for UCID to be used in a software engineering project. You are implementing it for the first time and would like to know certain things at the end of the project. First, you would like to know the overall software usability benefits derived from UCID. You would also like to know to what extent you really achieved the critical design goals for each information element. This is to convince yourself of the advantages of UCID as well as to get continued management support. Second, you would like to know the productivity levels achieved. Third, you would like to identify areas of quality and productivity where you failed so that you can establish meaningful goals for improvement.

To get answers to the above questions, you need to establish a *metrics program*. Define and use metrics in all UCID projects. It is a good idea to collect both quality and productivity, though you may want to get started with the former. When you have data collected from many UCID projects, you will have a *baseline*, which provides a series of measures that you can use to predict the quality, time, and cost of UCID activities and deliverables.

To see how your information elements fared, you can use user performance metrics and inspection indicators. *User performance metrics* can be based on CDGs. Examples are "time to retrieve information" and "time to use information." *Inspection indicators* can be based on CDGs, too. For example, you can have indicators relating to retrievability of information, such as "size of index" and "number of headings."

Pick and choose metrics wisely and use all the ones you select. If you are just starting a metrics program, you'll probably want to begin with quality. Along with the metrics, you should also collect important details such as number of people involved, experience and training of people, adherence to process, methods used, and use of software tools.

Prepare a *performance report* at the end of the UCID project. Record the metrics and other details you collected, project milestones (original vs. actual), and so forth. The purpose of the report is to record an assessment of the project so that you can learn what went right, what went wrong, how to build on the actions that led to success, and how to change the actions that caused problems. The report can be used to plan more effectively and to improve your UCID performance.

References

[1] Farkas, David K., and Steven. E. Poltrock, "Online Editing: Mark-up Models, and the Workplace Lives of Editors and Writers," *IEEE Trans. on Professional Communication*, Vol. 38, No. 2, June 1995.

[2] Baecker, et al., "The User-Centered Iterative Design of Collaborative Writing Software," *INTERCHI '93 Proceedings,* pp. 399-405, April 1993.

[3] Hackos, JoAnn T., *Managing Your Documentation Projects*, New York, NY: John Wiley & Sons, Inc., 1994.

5

Analyzing Users and Tasks

To be effective, all software usability components must be designed around tasks that users will perform—considering the users' knowledge, capabilities, limitations, and environment. The activity of getting to know users and tasks described in this chapter is, therefore, a prerequisite for designing the software usability components.

5.1 What is analysis?

Here is a quick, brief definition of user and task analysis. User analysis is the activity of getting to know the characteristics of people who will use the software system. Task analysis is the detailed study of tasks people perform manually or want to perform using the software system.

Analysis results are important for people who design the software usability components. In a user-centered design project, analysis will not be required for the design of information components alone, although this is the impression one gets from some technical writing books. For those who design the interaction components, knowing which group of users performs which tasks can, for example, help balance ease of learning and ease of use. Such knowledge can also indicate the need for a fast path and so forth. For technical writers who design the information components, analysis results help in the design of information architecture, specifications, and prototypes. It also provides inputs for writing the drafts of various support information elements.

Analysis is the first phase in the user-centered design process. Perhaps the only document available at this stage is a three-page product proposal. Typically, usability engineers perform the analysis activities in consultation with systems analysts and others. Technical writers should join this team early to get firsthand details about users and tasks. Once analysis results are available, they should be written in a format that can be used by all the designers. The results of user analysis appear as *user profiles* and the results of task analysis are summarized as *task descriptions*. Both appear as part of the usability plan. A brief summary can also appear in the UCID plan. Once user profiles and task descriptions are available, technical writers have much of the background material they need to get started with information design activities.

5.2   Why joint analysis?

Traditionally, technical writers have not contributed to the kind of task analysis practiced by the human-computer interaction discipline. But they have participated in user analysis. By joining usability engineers and others in user and task analysis efforts, technical writers not only get data for their work, but also contribute their own skills to the effort.

One of the important advantages of joint analysis is that technical writers get to know users firsthand. The primary reason why such direct contact is required is that technical writers are designing a part of the software system. They are designing interaction and support information—two of the four software usability components. Therefore, technical writers need user/task analysis data *almost* as much as other designers in the team need it. As designers, technical writers must proactively and directly seek to know about users. Without the knowledge of the people who will be using information, writers are likely to design and build information elements that do not fully meet users' information needs. With joint analysis, user profiles are based on facts, not opinions.

Then, there is the cost advantage. First, you cut costs as you eliminate the waste of developing more information than what users actually need. Second, the data you gather about users can be useful for future products aimed at the same groups of users. The cost is therefore spread over many projects. Finally, there is the money earned via "designing for users."

The advantages of joint analysis may be obvious. But there are still a couple of obstacles to direct involvement of technical writers. Hackos [1] says that technical writing managers admit the importance of analysis but discourage writers from directly performing it because of time and money constraints in their budget. Further, technical writers may not be actually trained to perform analysis activities, especially task analysis, which can be difficult and complex for them to do all by themselves. Fortunately, working jointly with analysis experts such as usability engineers helps them out here.

5.3   Getting to know the users

> Every step in understanding the users and in recognizing them as individuals whose outlook is different from the designer's own is likely to be a step closer to a successful design [2].

You should, as a designer, know your users. How does this knowledge help? With it, you can set your viewpoint at the users' eye level, include information they need, and cut out information they do not need. For example, by knowing which tasks are hard and which are easy for a group of users, you can then decide to provide more information for the harder tasks.

However, there is the challenge of user diversity. Groups of users have completely different information needs. Some need all the details, while a few others may choose not to use any information element at all. Some find information more easily using one design, while others may find the same design technique hard to use. Some prefer online Help, while some want to print the information in order to read it! Some want to learn everything before touching the keyboard, others want to jump in straight away. Then there are the international users to add to the diversity. Moreover, users change. Those who earlier preferred the mouse may now find the keyboard faster. Even users' attitudes and satisfaction levels change.

To handle this diversity, you really need to do three things: gather data about users, classify users into groups, and write profiles for each group.

What data about users should you gather? You need to know two types of characteristics about users. In the absence of better terms, let's call them general and specific characteristics. During user analysis, you will gather data about the specific characteristics of users. General and specific user characteristics are described in the following two sections. The subsequent sections describe the activities of classifying the data gathered and writing user profiles.

5.3.1 General characteristics

General characteristics of users are what you should know to successfully design any software. For example, you need to know that human beings are often subject to lapses of concentration. This is a general human trait. You probably cannot find via analysis which groups of users have this trait and which do not. Other examples of general characteristics are: response to external stimuli, problem-solving skills, learning skills, memory, motivation, emotion, prejudice, fear, and misjudgment. These are strengths and limitations that are common in human beings. For practical reasons, we cannot capture or measure these characteristics for every software engineering project. However, designers should keep them in mind while designing interaction and information components.

Let's look at learning styles, one of the general characteristics. We've seen in Chapter 1 that in order to quickly and easily use a software system users should have certain knowledge components. And that if they do not already have them, they must learn. To get users to learn, it helps to know how they learn. This knowledge helps us design a useful information architecture to start with. Brockman [3] describes learning styles based on whether users are aural, visual, or experiential. *Aural learners* like to listen to instructions rather than read instructions themselves. They prefer classroom training or having another person sit down beside them, guiding them as they work. They would probably best use the audio or video media. *Visual learners* want to have a clear picture before doing anything. They want to know what the software is, what it can do, and how it is used. They would best use printed support elements. *Experiential learners* like to learn by doing. They try something and if it does not work, they try something else. Perhaps they would best use online Help.

5.3.2 Specific characteristics

The second set of characteristics, the *specific characteristics,* are what you will gather perhaps for every software engineering project. These characteristics include the following:

- *Physical characteristics.* Gender may impact your graphic design, for example. Age can affect learning rate as well as preference for graphic design. You should also know details about disabilities such as low vision or hearing impairments.

- *Other characteristics.* Get to know users' education (what type and when), ability to read English, reading grade level, motivation, work style, culture, country, region, native language, and any expectations.

- *Application domain knowledge.* This is knowledge about an area, such as banking, that the software automates. Users, such as banking employees, may be required to possess banking knowledge. To get a clearer idea of users' application domain knowledge, you should get answers to questions such as the following: What is the level of knowledge and experience users have in their job domain? What are the job designation, roles, and responsibilities of each group of users? What job-related training have they taken (and how recently)? What are the tasks they will perform using the software? What role does the software play in their work?

- *Computer-use knowledge.* Do users know how to use the keyboard and mouse? Do they know how to use the operating system for which this software is being developed?

■ *User interface familiarity.* How are user objects and user actions implemented in *similar* software that users have used? What are the users' experience and skill levels?

■ *Computing environment.* There is a significant difference between using a windowing software on a 486 computer with 4 MB RAM and on a Pentium computer with 32 MB RAM. The difference is not speed alone, but can include graphics display capabilities, even users' motivations. What computing resources do users have?

■ *Use environment.* What about physical factors such as space, heat, light, and furniture? For example, is there enough space to keep manuals? What about social factors such as employee relations? And finally, what about work factors such as frequent interruptions, high noise levels, and pressure to quickly complete work? All these factors can impact error rates, user motivation, and so forth. You should make site visits to learn about these environmental characteristics.

5.3.3   Classification approaches

Let's look at a few ways of classifying groups of users:

■ Based on computer literacy;

■ Based on product-use skills;

■ Based on user roles.

While classifying, keep in mind that user expertise grows with increasing use of the software system. An expert in one task can be a novice in another. You may need to tailor the groups for your project. It is a good idea to give a percent for each group so that you can identify primary users and ensure that their needs are definitely met.

5.3.3.1   Based on computer literacy

Three categories of people are possible, based on level of computer literacy. *Lay people* have never touched a computer keyboard. Anything about computers is new to them, including computer terminology. The trend is that the number of people who want to use computers is continuously growing. The *computer literate* have used a computer and are familiar with a subset of computer terminology. *Computer professionals* are programmers, analysts, designers, and information systems managers.

5.3.3.2 Based on product-use skills

The groups are novice, intermediate, and expert.

Novices   Novices are people new to the software system. You should get novices to appreciate how the software system helps them do their work. They will need information to learn to use the software. And they are likely to have time allocated for learning. Usually, novices have a set of questions but lack the vocabulary to express their questions. Answering these questions can largely help meet their information requirements. Novices only use a few of the software system's capabilities. Rather than power and performance, they look for defaults and templates available in the software system.

Intermediates   These are the "I have used it before" people. Intermediates are actually novices after some experience. They may use defaults and templates, but start looking for ways to pack power into their tasks. Intermediates use many features of the software without referring to support information.

Experts   Experts are users who thoroughly understand how to use the software. Experts know how the software system is organized and how it works. They exploit the features and flexibility of the software system to maximize their performance. They combine options for power and know the impact of input parameters on the output. For this, experts need information on advanced topics and expect every option to be described fully and in detail. They may want a quick reference card to remind themselves of procedures or parameters. To experts, menus are obstacles. They demand shortcuts and a fast response time.

5.3.3.3 Based on user roles

This approach is appropriate if many support tasks are performed by people with distinct responsibilities. See Table 5.1 for an example.

Table 5.1
User Roles (an Example)

User Role	User Tasks
Product Administrator	Set up customers Set up units in which customers will be billed Set up pricing policies Perform budgeting and forecasting
Managers	Print and use reports
Data Entry Clerks	Type billing details

5.3.3.4   Occasional and transfer users

Any of the users described in the preceding sections can also be occasional or transfer users. Here is a brief description.

Occasional users   These are people who come back to the software system after a lapse of time. Occasional users only accept a limited amount of training. They do not want to know much about the software system. They make frequent errors. And because of infrequent use, occasional users may want to look up a forgotten command or procedure.

Transfer users   These are users experienced with a similar product or the previous version of the same product. Transfer users are now trying to learn a similar software system—that is, they are trying to transfer what they already know to a new system. They may need to first unlearn what they know and then learn the new user interface. They are likely to start with tutorials and guides. The trend is that the number of transfer users is increasing because more people today are going for better software systems at home as well as at work.

5.3.4   Writing the user profiles

Write user profiles once user analysis data is gathered and users have been categorized into groups. A user profile is a description of a particular category of users. The details should be limited to what is potentially helpful in designing the software usability components. Highlight any differences that could affect the product or information use. Specify the proportion of total user population each category constitutes. See Table 5.2 for an example. User profiles should be documented in the usability plan. Brief summaries can also appear in documents such as product proposals and functional specifications.

Table 5.2
User Profiles (an Example)

User Role	User Tasks	User Characteristics
Product Administrator	Set up customers Set up units in which customers will be billed Set up pricing policies Perform budgeting and forecasting	2.7 years application domain experience 31% hold bachelors or masters degree Frequently use product Varied information systems knowledge
Managers	Print and use reports	2.4 years of application domain experience 55% hold bachelors or masters degree Minimal use of system Medium to minimal information systems knowledge Have used similar products for reports generation
Data Entry Clerks	Type billing details	6 months job experience 70% know typing High school education Little information systems knowledge Frequent but limited use of system Not knowledgeable about the purpose of the product

5.4   Getting to know the tasks

As we have seen in Chapter 1, tasks can be end-use tasks or support tasks. End-use tasks directly contribute to user's work, achieving the purpose for which the product is used. They need to be understood from the users' perspective. Tasks are of central interest in user-centered design.

Human factors studies have indicated the need for a task-based approach to designing software for usability. To design task-based software, designers need to know about the users' tasks. This knowledge is gained via task analysis. In fact, if task analysis is done well, the resulting data may itself indicate the design approach that is more likely to succeed. Further, task analysis helps the software project manager better scope the project.

Despite its clear need, task analysis is not very clearly understood. People find it difficult because it involves the analysis of tasks in terms of human behavior. This can be complex and tedious.

5.4.1   What is task analysis?

There exists at least three views to task analysis. One view is from traditional programming, where it is called systems analysis. The second view is from technical writing. The third view is from the discipline of *human-computer interaction* (HCI). In your UCID project, you will follow a joint analysis approach.

The traditional analysis view   Traditional systems analysis techniques have been effective for specifying the functional requirements for new software, but they place little or no emphasis on the usability of the resulting software.

The technical writing view   The technical writing profession appears to view task analysis more as a technique for deriving methods of organizing and writing manuals than as a way of looking at and describing the tasks themselves. The assumption is that the software has already been designed. Technical writers just "analyze" the existing software to determine the tasks, order of tasks, and so forth. During this "analysis," they ask questions like: What are the steps? Who does it? What are the prerequisite tasks? When is a good time to perform these tasks? The answers help the technical writer to better organize and write a manual.

The HCI view   There are certain things common between the systems analysis view and the HCI view of task analysis. Both attempt to define user requirements in terms of functionality. Both use a hierarchical decomposition of the tasks people perform or want to perform. But the HCI view is different in that it analyzes tasks in terms of human behavior.

Hierarchical task analysis (HTA) is a method directed at decomposing a task into the necessary goals and subgoals, and procedures for achieving those goals. It requires the study of actual users to form a detailed model of the task. The strength of HTA is in its empirical content [2].

The problem with hierarchical decomposition is that it can be hard both for analysts to analyze and for designers to use. To minimize the complexity, the trend is to seek contextualized, user-centered representations of tasks via, for example, scenarios of typical work. Scenarios make task analysis data usable. Write scenarios that illustrate typical sequences of actions with real data. Include user exceptions that typically disrupt real-life tasks. Examples of exceptions are typing errors or interruptions like switching to another, higher priority task [4].

Scenarios are real and detailed examples of things users actually need to do. Lewis and Rieman [5], while talking about task analysis for a traffic modeling system, mention the following as one of the tasks they developed: "Change the speed limit on Canyon Boulevard Eastbound between Arapahoe and 9th. Calculate projected traffic flows on Arapahoe West of 6th assuming Canyon speeds between 25 and 55 in increments of 5 mph." This example describes what the user wants to do, but not how. It is very specific, even specifying street names. It is also complete, as it covers both changing the speed limit and calculating the projections for Arapahoe. Such a description should also specify who will perform the tasks.

Scenarios also nicely complement the requirements produced by traditional software engineering. For a document development software package, for example, a scenario might be to produce the book, *User-Centered Information Design for Improved Software Usability* by Pradeep Henry. This scenario supplements the detailed partial tasks collected in traditional requirements analysis, which might include things such as "key in text," "check spelling," and "print the document." The traditional approach helps to ensure that all important functions of the system are recognized, while the representative tasks in the task-centered approach provide an integrated picture of those functions working together.

Joint analysis in UCID   In UCID, both the HCI view and the technical writing view are important. The HCI view is jointly executed early during the UCID analysis phase by usability engineers and technical writers. Later, technical writers carry out the technical writing view for designing the information architecture.

5.4.2   Writing the task descriptions

Record task descriptions in the usability plan. Update them later to include the details required for the technical writing view of task analysis. Some of the questions the task description should answer are:

- What is the task?

- What are the subtasks?

- Who performs it?

- How often is it performed?

- Where and with what computing resources is it performed?

The updated version will also answer questions specific to the software being developed: How does the user know when the task is complete? How do users recover from errors? See Table 5.3 for a simple example.

5.5   Gathering data

In carrying out analysis, three questions need to be answered:

- What data should be gathered?

- Where can it be obtained?

- How can it be obtained?

Table 5.3
Task Description: A Simple Example

Installation	This support task involves installing the product on the host computer using the installation utility. Specifics of the task include: ■ Preparing for installation on the host computer ■ Installing the product using SMP/E ■ Tuning the product to affect performance
Administration	This support task involves controlling how the product is used by end users, and defining and tracking business policies to meet the planned processing goals of the organization. Specifics of this task include: ■ Defining and tracking the billing and pricing policies ■ Authorizing users to perform selected end-use tasks ■ Defining component-specific and non-English profiles ■ Customizing component-specific information
End-use	End-use tasks involve the use of the product for its intended purpose. The end-use tasks are: ■ Defining new customers ■ Defining units in which customers will be billed ■ Price setting ■ Billing ■ Forecasting ■ Budgeting

Irrespective of the method of gathering data, you should have a check-list or questionnaire that will remind you to get all the answers you need. There are various ways to get data. Some are described here.

■ Make customer site visits. This method should be your top priority. Talk to customers face to face at their workplace. There is no better alternative to studying users in their natural work environment. You get a better understanding of their needs, and the challenges and difficulties they face at their actual workplace.

■ Participate at user groups and customer training sessions.

- Consider hiring a market research firm. This is for user analysis only—to make a demographic study of the product's market.

- Conduct surveys by mail or phone. Unfortunately, questionnaire surveys have been found to be the least productive way to get data. Telephone surveys can be cost-effective if carefully planned. Careful planning includes deciding on the sample and who you will call.

- Contact internal sources, especially the departments that have direct contact with users. Examples of such departments are customer support, marketing, training, as well as planners and developers of similar products and support information. If the software system is for an in-house user group, it is easier to "catch" them. You can also get data about these users from your human resources department.

- Read all you can. Read marketing and contractual documents (RFP, project proposal, etc.), technical documentation, and manuals for similar products, projects, or clients. Read user survey reports. Also, check out trade publications of the industry in which users work.

References

[1] Hackos, JoAnn T., *Managing Documentation Projects*, New York: John Wiley & Sons, Inc., 1994.

[2] Johnson, P., *Human-Computer Interaction: Psychology, Task Analysis and Software Engineering*, London, U.K.: McGraw-Hill Book Company, 1992.

[3] Brockmann, John R., *Writing Better Computer User Documentation: From Paper to Online*, New York, NY: John Wiley & Sons, 1992.

[4] Curry, M. B., et al., "Summarising Task Analysis for Task-Based Design," INTERCHI '93 Adjunct Proceedings, 1993, pp.45-46.

[5] Lewis, Clyaton, and John Rieman, *Task-Centered User Interface Design: A Practical Introduction*, Boulder, CO: Shareware Book, 1994.

6

Goal Setting for Software Usability and Information Quality

B OTH USER-CENTERED DESIGN and UCID approaches ensure quality through the following quality process:

1. Set goals;

2. Design with goals in mind;

3. Evaluate for achievement of goals;

4. Improve;

5. Iterate.

This chapter is about step 1 in the quality process. There are two objectives for the chapter. One is to define software usability goals through interdisciplinary collaboration. The other is to define information quality goals that contribute to maximizing the software usability goals. Toward defining information quality goals, we will learn how information is applied by users. We will also learn how to use the information use model to come up with a set of CDGs.

6.1   Setting software usability goals

> Without specifying the degree of ease or pinning down the claims in a more concrete way, "easy to use" is just puffery [1].

Easy can be defined as "requiring little effort" and "posing little or no difficulty." Easy is not very useful as a software usability goal statement because it is a relative term. What is easy for one user group can be hard for another. So we need more concrete, preferably measurable goals. The user-centered design approach largely drives design with measurable goals. An example of a measurable software usability goal is "Average transaction time will be less than 20 seconds for clerical personnel who have used the software system for one week."

6.1.1   Why define software usability goals?

There are may good reasons why you should set software usability goals. Goal setting guides the design effort. When you have clearly defined goals, making trade-offs among design alternatives is easier. You can evaluate each design against the goals and determine the one that best achieves them. Goal setting supports an iterative process, which helps you create a good design. Moreover, it tells you when you can stop iteration and obtain sign-off from users and responsible people in your project or organization.

6.1.2   Usability goal-setting process

Usability goal setting has traditionally been done by usability engineers. Why does UCID call for a collaborative effort with technical writers? As described in Chapter 1, two out of the four software usability components are information components: interaction information and support information. In UCID—unlike in traditional approaches—technical writers design interaction information, too. Therefore, a collaborative effort is required to specify the goals for software usability. Of course, inputs, at least via reviews, should also come from programmers, marketing specialists, and project and corporate management.

Here are the activities related to specifying software usability goals:

- Establish a baseline;
- Specify goals;
- Prioritize goals.

Establish a baseline   A baseline is a repository of empirical data about the usability of the existing or a competing software system. It is the foundation for setting informed usability goals. Baseline data is also useful for making comparative usability assessments.

Specify goals   During the UCID analysis phase, you should define software usability goals and document them in the usability plan. Goals should be defined based on a knowledge of baseline data, market requirements, user profiles, project requirements, and corporate objectives.

You might have in mind general goals such as "easy to use" and "fast to use." However, it is important that you define specific goals such as the following:

1. Average time for users to complete a task is less than 8 minutes. Users have no experience using the software system.

2. Number of errors made by users while performing a task is less than three. None of these errors cause data to be changed or lost.

3. At least 75% of users should rate the following as good or very good. The five-point scale ranges from 1 = very good to 5 = very poor.

    - Overall satisfaction;
    - Ease of navigation;
    - Quickness of error recovery.

Typically, software usability goals are either user performance-related or user preference-related. Examples 1 and 2 above are performance-related, whereas example 3 is preference-related. Performance-related goals are measurable goals. Here, the goal statement typically consists of a metric such as "time to complete a task" and a numeric value, such as in example 1.

User performance data (or metrics) refers to objective, measurable indices of usability. The measures are often in terms of rates (percent), time measures, or counts. Such data is usually collected via testing. The following are examples of user performance data:

- Time to complete a task;

- Time spent in recovering from errors;

- Number of tasks performed correctly;

- Number of assists required;

- Number of times support information was retrieved to complete a task;

- Number of errors;

- Number of times user is misled;

- Number of times user loses control of the software system;

- Percent of users who complete a task without assistance;

- Percent of favorable/unfavorable user comments;

- Percent of errors;

- Percent of users who complete without a single error.

User preference data refers to subjective data of users' opinions on the usability of the software system. It is usually collected through a questionnaire. An example of preference data you might want to collect is "How easy is it to navigate?" The response could range from "very easy" to "very difficult."

Decide on the various performance and preference data you should collect. For all performance data and for some preference data, you will want to specify a numeric value. Typically, the numeric value is the target level you want to achieve. In other words, it is the software usability goal. Here are some ways to determine target levels:

- Based on the existing software system;

- Based on a competing software system;

- Based on performing a task manually;

- Based on a prototype of the software system being developed;

- Based on earlier performance using the software system—by the same users.

In addition, you might want to specify numeric values for current level and minimum level. Make sure that the minimum level is better than the current level or baseline. See Table 6.1.

Prioritize goals Assign priorities to indicate the relative importance of each goal. This will help you make appropriate trade-offs between software usability goals and other organizational or project requirements.

6.2 Setting information quality goals

Information is high on quality only when it contributes to overall usability of the software system. Goal setting, described in this section, supports this objective.

Table 6.1
Assigning Values

Data (metric) to Collect	Minimum Level	Target Level (or software usability goal)
Percent of errors	Equal to (name of competing product)	50% better

Goal setting for information quality has important advantages. It directs writers to produce information elements with very specific goals in mind. Consider this goal: "Time to find and display specific information should be less than 20 seconds." When you have a goal like that, all design, writing, and formatting decisions are focused on the objective of reducing the time spent using the information element.

Once the information architecture is ready, a senior technical writer/manager should specify information quality goals in consultation with usability engineers. Review inputs should come from programmers, marketing specialists, and project and corporate management.

Here are the things you should do to define information quality goals:

1. Visualize the information use model for each information element;

2. Know the information design goals;

3. Define critical design goals;

4. Define information quality goals.

6.2.1   Information use model and information design goals

Information quality goals are based on *critical design goals* (see next section) that you will define for each information element. The information elements themselves are identified by (1) visualizing how the software will be used and (2) looking at users' requirements concerning the need for, and use of, information. Toward that end, you need to think of the tasks users will perform, the task scenarios, the task environment, users' characteristics, their information needs, and how they will use each information element.

When users need information, the design of the information elements should generally meet the following requirements concerning its use:

1. The required information is available;

2. The information can be quickly and easily obtained;

3. The information can be quickly and easily read;

4. The information can be quickly and easily understood;

5. The information can be quickly and easily used;

6. The whole experience is pleasant;

7. The information works.

Let's translate each of these use requirements into a single-word design goal. On translation, point 2 for example becomes "retrievability," which is a well-known information design goal. See Table 6.2 for a list of information design goals. To aid recall, let's call them by the acronym ARRCUPA.

The information use model described above in terms of seven use requirements is a general model. For each information element you provide, it may have to be slightly tailored. The model depends on how you expect the information element in question to be used.

Table 6.2
Information Use Model and Information Design Goals

Requirements	Information Design Goals (ARRCUPA)
The required information is available	Available
The information can be quickly and easily obtained	Retrievable
The information can be quickly and easily read	Readable
The information can be quickly and easily understood	Clear
The information can be quickly and easily used	Usable
The whole experience is pleasant	Pleasant
The information works	Accurate

6.2.2   Critical design goals

Information quality goals are based on CDGs. CDGs are critical success factors for the information elements you provide. For every *significantly unique* information element you create, you should define a set of CDGs. The set can be the same for two or more information elements if these elements are similar and are used in a similar way.

You should derive CDGs primarily based on the information use model and the information design goals for the information element. You should also consider two other things: the design problems common in the information element in question, and the overall objective of improving software usability. Record CDGs in the UCID plan.

Since the information use model could be different for each information element, the associated information design goals could also be different. Consider the information design goal "clear." Clarity is achieved via subgoals such as user viewpoint writing, conversational style, conciseness, consistency, and examples. In the case of messages, for example, "user viewpoint writing," a subgoal of "clarity" is critical for the success of the message text. Such information design goals that you tailor for an information element are the CDGs for that element. In the case of our example just described, you'd define "user viewpoint writing" as one of the CDGs, rather than a more general "clear." This book defines CDGs for labels, messages, and online Help in Chapters 9, 10, and 11, respectively. You should also define CDGs for other information elements you may provide.

You should design information elements keeping their CDGs in mind. CDGs keep your efforts focused on things most important to users, things that help them complete their tasks quickly and easily. With CDGs, information evaluation is finally made meaningful: you evaluate each information element against its CDGs. This help reviewers, too, since they now know what to look for. The bottom-line is that CDGs help you deliver information that users perceive and realize as "making it easier."

6.2.3   Information quality goals

Though specific to your information element, CDGs are still general goals. Just saying that an information element should be, say, *retrievable* is not useful enough. You need to come up with more specific, measurable goals called *information quality goals*. For retrievability, the information quality goal could be "Time to find specific information should not exceed 60 seconds."

You should have separate sets of information quality goals for each information element. Consider defining one information quality goal for every CDG. The procedure to define information quality goals is the same as that earlier described for defining software usability goals. While defining information quality goals, you should consider software usability goals, market requirements, project requirements, and corporate objectives. Record your information quality goals in the UCID plan.

Just as in the case of software usability goal setting, identify both user performance data (or metrics) and user preference data that you will collect. Examples of user performance data are "Number of users who successfully completed the task after reading the information" and "Number of times user looked in the wrong place for information." An example of user preference data you could collect is "Are labels understandable?" Finally, assign numeric values for various levels of quality, such as target level and minimum level.

Reference

[1] Potosnak, K., "Setting objectives for measurably better software," *IEEE Software*, March 1988.

7

Designing the Information Architecture

A COOK IS SUPPOSED to have said: "The flour itself does not taste good, nor does the baking powder, nor the shortening, nor the other ingredients. However, when I mix them all together and put them in the oven, they come out just right for biscuits." This chapter is about determining the right combination of information elements that will maximize the usability of the software system. The activity is called integration, and the resulting design is called information architecture.

7.1   What is information architecture?

The information architecture is the result of the integrated approach to information design. It is the blueprint for maximizing software usability via the integrated design of labels, messages, online support elements, and printed support elements (see Figure 7.1). The information architecture identifies all the information elements users need and expect, describing each in terms of content, media, and form—all based on the needs and expectations of users.

The information architecture helps cut time and cost by ensuring that big design errors are found and corrected early. You know it is much easier to cut four chapters from an information architecture than to add three pages to a final draft. Just as the architecture plan for a building project coordinates the activities of many workers, the information architecture coordinates the activities of a number of writers. For the technical writing team, it makes information development a more professional endeavor.

7.2 Integration: what is it, why is it important?

For real usability, the entire user interface, including all aspects of every interaction with the user, should be designed as a single, integrated element of the system [1].

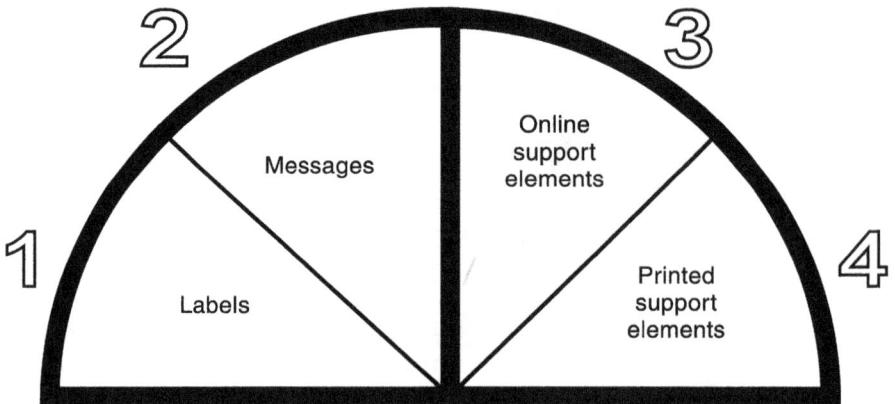

Figure 7.1 Four-tier information architecture.

If all the information elements you produce are good "documents" individually, you will maximize software usability, right? While talking about the reasons for hard-to-use systems, Jeffrey Rubin [2] says, "... this (non-integrated) approach is futile, since it matters little if each component is usable within itself. Only if the components work well together will the product be viewed as usable and meeting the user's needs."

Take for example a two-color, nicely produced user manual that has *all* the information about your software system. Though it also has that small piece of detail that a user is looking for, retrieving that detail will take a longer time because the manual is bulky, probably with many pages of unnecessary or inappropriately placed information. This will only adversely affect software usability. Some of the information in that manual should have been dropped or covered in another appropriate information element.

As described in Chapter 1, the usability components of a software system are user objects, user actions, interaction information, and support information. Gould and Lewis [3] say, "Unfortunately, these interacting pieces are usually designed separately. Definers, designers, implementers, application writers, and manual writers constitute large groups in themselves, and are often separated by geography or organization.... It appears that superior quality can be attained only when the entire user interface, including software, manuals, etc., can be designed by a single group, in a way that reflects users' needs, and then evaluated and tuned as an integrated whole."

What's critical, therefore, is integration. Integration is the software usability-driven approach to identifying all the information elements users need and expect, and designing, evaluating, and improving them as an integrated whole. In this definition, "identifying" implies deciding on the content, media, and form that users need and expect from each information element. And, "all the information elements" covers labels, messages, online support elements, and printed support elements.

Integration might appear obvious. But, do software houses integrate interaction and support information? Do they even integrate printed and online support elements? Here is what technical communicators have expressed about the process of developing printed and online information [3]:

- *Online documentation should be written after printed documentation.* The technical communicators' reasoning is that the online document is a subset of printed information. They also say that printed information contains many different structures such as introduction, procedures, examples, glossary, and reference. The problem with this approach is that online information is not designed specifically for online use. For example, online text taken from a printed user's guide may not provide the quick how-to information users expect while working online.

- *Online and printed information should be written independently.* The technical communicators say that the same text should not be used to serve different purposes. Therefore style, formatting, and retrieval techniques for online text will be different. Even in big companies, online Help, for instance, is often planned and written separate from the printed manual—often in another city. The problems with this approach are as follows:

  Information provided in, say, online Help is repeated in, say, a printed manual when there is clearly no such requirement from the users' viewpoint.

  Worse, if the Help (or manual) writer incorrectly assumes that the writer for the other media will cover certain topics, critical information could be left out.

  Error message might say that how-to information is available by pressing F1. When users press F1, all they get is what-went-wrong information.

- *Printed and online information should be generated simultaneously from a common database.* The technical communicators say that printed and online documents contain the same information and that this ensures consistency.

Here are the reasons why you should apply the integrated approach:

- Ensures that users have all the information they require—in the media and form they need or expect;

- Eliminates information users do not need or expect;

- Eliminates unnecessary repetition of information (note that repetition may be required from the user's viewpoint);

- Eliminates user dissatisfaction resulting from the above;

- Changes the user perception that using support information is an unnecessary diversion;

- Often results in smaller information elements with improved retrievability.

Because of interdisciplinary collaboration in UCID, when you take the integrated approach within a user-centered design process, you will have integrated information elements and, more important, the information elements will be integrated with the rest of the software usability components. Users will perceive the overall software system as one single usable system.

7.3   Information architecture: design process

The information architecture should be designed by the technical writing manager or senior technical writers in consultation with the usability engineer and the programming team. The initial information architecture is designed during the UCID high-level design phase. It should be improved based on task (user interface) complexity data, which may be available later in the project.

Integration clearly is the way to maximize software usability. Unfortunately, there is no single right mix of information elements—a predetermined set—that you can provide for every software system. For example, among the three software products of Netscape Enterprise Server, Microsoft Word, and IBM Interactive Storage Management Facility, everything is different. The user interface is different. User groups are different. Information requirements are different. Obviously, the mix of information elements for each of these products is different. Similarly, you should determine the specific information elements required to support the information needs and expectations of your product users.

Nor can we define a comprehensive "IF...THEN..." formula. For example, I cannot say, "If you provide online Help for error messages, you don't need to provide a Messages section in the printed guide" because that depends on many factors that only the project technical writers know.

Therefore, what I provide here is only a framework to get you started on the right track.

The question you are trying to answer is, "What information elements should be provided?" In other words, what sets of topics should be separately packaged for which media? To answer this question, you should know the following:

1. *Users' information requirements.* You should know users, tasks, and task (user interface) complexity.

2. *Media characteristics.* You should know which media is best for what type of information and what user situation.

3. *Project-specific requirements.* You should be aware of things like resource and schedule limitations that are specific to the project.

Armed with that knowledge, you should do the following:

1. Visualize the use of the software system to determine what information users will need and *when*.

2. Pack a set of topics into each information element. For this, you should also consider various packaging approaches, such as task-orientation described later in this chapter.

3. Title each information element.

4. Document the information architecture. List all the information elements that you should provide, describe what each element contains, and show how each element fits into the whole. You can place these details in the UCID plan.

7.4 Know information requirements

Information elements provided should contain just the information users need and expect. Exclusion of some information may affect the users' task performance. Presenting information that is not required is a waste of time and effort. Moreover, it worsens retrievability and makes manuals bulky.

To find what information users need, you need to know the users, tasks they perform, and the complexity of the user interface.

7.4.1 Users and their tasks

The more you know about users and tasks, the more you will know about the information required. See Chapter 5, which is devoted to this topic.

One of the things you should do to determine information requirements is construct a matrix, where you list the tasks against the user groups that will perform the tasks. The simple user/task matrix in Table 7.1 indicates the possibility of each user group performing a task. The ranking is on a 0 to 5 scale, where 0 indicates that there is no possibility. Use this matrix to identify the most important user-software interactions.

In another user/task matrix (Table 7.2), you could show which user group will use which information elements to perform what tasks. It might be useful to summarize here the results of a survey [4]. The survey asked experienced users and writers of online information to rank their preference of four "instructional" elements on a scale of 1 (most preferred) to 6 (least preferred). Here are the results:

Table 7.1
Matrix Showing the Possibility of a User Group Performing a Task

User/Task	Clerks	System Administrators
Installation	2	5
End Use	5	0

Table 7.2

Matrix Showing Who Uses Which Elements to Perform What Tasks

User/Task	Clerks	System Administrators
Installation	Labels, messages, user's guide	Labels, messages
End Use	Labels, messages, online Help, user's guide	None

- online tutorial (average 2.58);

- classroom course (average 2.65);

- hardcopy tutorial (average 3.0);

- trial and error (average 4.0).

7.4.2 Task (or user interface) complexity

This is about how users will perform the tasks as implemented in a particular software system. Knowing task complexity is an excellent way to know what types of information users will need and *when*. You will, however, need at least some idea of the look and feel of the user interface, which is unlikely to be available at the start of the UCID high-level design phase. Iteration of the information architecture should therefore continue, at least until some user interface is available.

To understand task complexity, you could try a technique described by Doris [5]. Using a list of 10 questions, Doris set up a difficulty index to describe the tasks users can perform with a software system.

1. Does this task require a single key response?

2. Does this task require the use of more than one key?

3. Does this response require synchronized operation of two or more keys?

4. Does this task response consist of a sequence of keystrokes?

5. Will users need to reposition the cursor?

6. Does this task require numbered steps to explain it?

7. Does this task response result in one or more new screens?

8. Does this task require a typed response?

9. Does this task require a choice from a menu?

10. Does completion of this task leave users on a new screen?

An analysis based on a difficulty index can indicate user interface complexity. In the context of the example software described, Doris remarks that the easiest task is "Deleting a character to the left," which required only one keystroke. The most difficult task is "Selecting a specific page to print," which requires the use of five key responses, involves three or more steps, results in two new screens, requires two typed responses and three menu choices, and results in users being left on a new screen.

Such indications of user interface complexity help you in designing the information architecture as well as in writing the information elements.

7.5 Know media characteristics

The medium of information delivery is a decision you make at the time of designing the information architecture. To make that decision, you must first understand the particular media where users will need or expect certain types of information. You should also know what kind of information would better go online and what would better remain on paper—from the users' viewpoint.

Print and *online* are the media most often used for communicating information. In the following sections, we will look at the effectiveness of online and print media in terms of retrievability, usability, and readability.

The *World Wide Web* is a new media for providing support information. Although information appears on the users' screens, I'm not discussing Web support information under online media because of the Web's own unique characteristics. Today, you should consider Web-based information elements for your software products. The possibilities are enormous, including multimedia capabilities and pointers to useful

Web sites and newsgroups. Cost savings are possible for the software house that cuts down on printed information. Disadvantages with Web support information are the need for Internet access and slower access to information.

Video is another support information media. It is primarily used for presale demonstration and tutorials. The advantage is that it is fairly well suited for demonstration. The disadvantage is that it demands a TV and VCR.

The *audio (tape)* media is used, but rarely. Its obvious advantage over online media is the absence of screen clutter! The disadvantage is that it demands a cassette player.

7.5.1 The print media

Even after the "invasion" of the online media, print media is still used for its many advantages. Examples of printed information elements are user's guides, reference cards, and instruction foldouts.

Types of information better suited for the print media include information that:

- Is "a lot of text;"

- Contains concepts;

- Demands careful or detailed reading;

- Must be available when a computer is unavailable (e.g., installation and crash recovery information);

- Users may want to take home;

- Is legal, such as copyright, disclaimer, license, warranty, and safety information.

7.5.1.1 How effective is print media

How good is print media? Let's review it in terms of retrievability, usability, and readability attributes.

Retrievability Compared to the online media, retrievability with print is generally slower. However, there are many techniques you can use to maximize print retrievability. Chapter 13 describes retrievability techniques, which you can consider for your printed information elements.

Usability Printed support elements are difficult to make available to everyone. Often, users need to walk up to the bookshelf or library. Printed elements can get loaned, misplaced, or lost. They take up desk space. One advantage is that most users find it easy to perform tasks online as they read instructions from a printed element. Finally, there is the advantage of portability.

Readability If designed carefully, printed information elements can have excellent readability. Here are certain factors that contribute. Resolution is excellent, with typeset pages having a typical resolution of 1,200 to 2,400 dots per inch. You can also use good graphic design to enhance readability. Size is a contributing factor. Size varies, but is usually small enough so that two full pages can been seen in one glance. Shape is another factor. It is predominantly portrait.

From the software house's viewpoint, one disadvantage with print is the production process it demands. A separate, time-consuming, and expensive production process is required through an external agency.

7.5.2   The online media

The online media is the computer screen, where support information is displayed. The information element is often "physically" part of the software system itself. Information better suited for online media include quick "how to" information. Examples of online information elements are Help, tutorial, and cue cards.

7.5.2.1   How effective is online media?

At information design seminars I give, when I ask participants which media is better, most of them would put up their hands for online. But when I get down to specific questions such as, "Is it easier to read on the screen?" they start thinking again. In fact, there are people—like me—who would rather *print* it and read. Just like print, online has its advantages and disadvantages.

Retrievability If designed carefully, online retrievability is clearly superior to print retrievability. Users can retrieve and display information with a keystroke. The best thing about online retrieval is that even if you have a large amount of information, users do not perceive going from one "manual" to another. Hypertext and other techniques make retrieval a breeze.

Usability One of the best things about online media is the fact that users perceive online information as the software system explaining itself. In contrast, a printed manual is physically separate from the system and, therefore, users might think of it as a hindrance to the task they are performing with the system.

   The online ability to display context-sensitive information has no parallel in the print media. Users can get just that piece of information they need at the moment—with one keystroke. Depending on the user's task context and his or her earlier performance, you can even automatically display contextual information.

   Users do not perceive the "bulkiness" of online information. Only what they request appears on their screen. Moreover, removing the information is usually quick and easy.

   Users, however, have complained about online usability in terms of them having to turn information "off and on" to continue using the instructions. This problem is eliminated in many new software systems, where a small window containing the instruction always stays while users read and apply the information to their application task.

   Bad design can create problems. If the user interface for an online support element (e.g. Help) is significantly different from that of the user interface for the software system, users have to learn how to use that element. This defeats the purpose of providing Help or other online support elements. Bad design can also cover or clutter users' working screens.

Readability Studies have shown that online readability is a loser against print readability. Resolution, the spacing between pixels on the screen, is not yet as good as on a printed page. Moreover, users cannot easily adjust the reading distance. Whereas most paper-based graphics can be shown on the screen, not everyone's computer will have the required resources, such as high-resolution monitors. The bottom line is that on-screen reading is 20% to 30% slower. For many users, online reading can result in eyestrain and headache. Online readability may be improved in the future when technology advances and become affordable.

7.5.2.2   Why developers think online is better
Here are some reasons why developers prefer developing information for the online medium.

Easier to produce and distribute   For producing online support elements, you don't have to use an external agency, whereas for printed support elements, you will go through a print shop. Of course, online information elements have to be put on a diskette or CD-ROM. However, this duplication usually happens alongside—and the same way as—the rest of the software system. In general, online information is also less expensive to duplicate, store, and distribute.

Better acceptance   Software engineers generally accept online technical writers more readily as fellow team members. Online writers themselves accept and enjoy the technical writing profession better because online information is "physically" part of the software system and because of the various technical procedures, such as testing, that is involved.

7.6   Know project-specific requirements

We may have a design that is excellent from the users' viewpoint. From a business viewpoint, however, the solution may be infeasible. It is important that we strike a balance between an excellent solution and an infeasible solution by understanding the realities of the business environment in which the software engineering project exists. Some of the business issues you should analyze are as follows:

- *Skills*. For example, is creative talent available for animation design?
- *Tools*. Are tools available for prototyping an online tutorial?

- *Cost*. Will this expensive tool pay off in the long term? Is the budget too low to produce a complex set of information elements?

- *Schedule*. Software houses are always looking for ways to reduce time to market. Is there time to learn the new tool?

7.7   Package information elements

Decide what will go into labels. Also, decide what types of information will go into different message types, such as informative messages and warning messages. For packaging of support information elements, consider the following approaches.

7.7.1   Packaging approaches

Three standard ways of packaging information are described here. They are task-orientation, role-orientation, and software-orientation. Books that describe these approaches recommend task-orientation alone. However, you should consider all three approaches to see which best meet users' information requirements. Note that here we are only talking about high-level packaging of information into elements. Low-level design and writing should always be task-oriented, irrespective of the packaging approach you implement. This is simply because users are performing *tasks* with the software.

Task orientation   The information to be provided is based on an analysis of the use of the software and is limited to what users need to perform specific end-use and support tasks. Information that is not required to perform tasks is useless information and is not provided.

Role orientation   If you use this approach, you will have each information element targeted at each user group. For example, you might have a system administrator's guide, a user's guide, and so on. You should consider this approach if roles such as system administrator are clearly defined in your customer organizations.

Software orientation If you use this approach, you will follow the software system's structure consisting of modules or functions. Here, you might have an information element each for a major function of the software system.

7.7.2 Some tips to consider

Here are some tips you should consider:

- Consider size. Avoid packaging a printed element of over 300 pages or less than 20 pages. A large reference is not too bad because users are just looking for a specific topic (in contrast, users may read an entire chapter of a guide).

- Keep all guide information together, but not necessarily in an exclusive information element. Depending on expected size and use, you can put guide and reference information together in a single information element.

- Avoid unnecessary repetition. If you want three manuals, all of which have, say, over 30 pages of common information, put that information in a single common manual.

- Avoid requiring users to jump from one information element to another.

- For mainframe software systems, users may be clearly defined as system administrators, operators, and so on. Here, you probably should not put system administration and operation information together in a single manual—especially if users have the option to order manuals.

- If your software system has a function that lets users make extensive changes to the software's operation, you may want to separately place the description of that function so that it is only available to "privileged" users.

7.7.3 Give it a title

The goal of the title for an information element is improved retrievability. The title can include some or all of these:

- Product name;

- Content;

- Approach;

- Users;

- Purpose;

- Company name.

Many software houses have their own rules for titling manuals. IBM mainframe manuals typically have the structure shown in Figure 7.2.

Normally, companies strive for consistency in titling. However, it is more important to evaluate titles for their retrievability.

7.8   Information architecture: the OS/2 Warp example

The example given here is a set of online and printed support information elements for an IBM product. It does not cover the interaction information.

The product's printed information element, *User's Guide to OS/2 Warp* [6], provides an interesting road map to the online information elements users may use. Let us first take a look at the printed element itself.

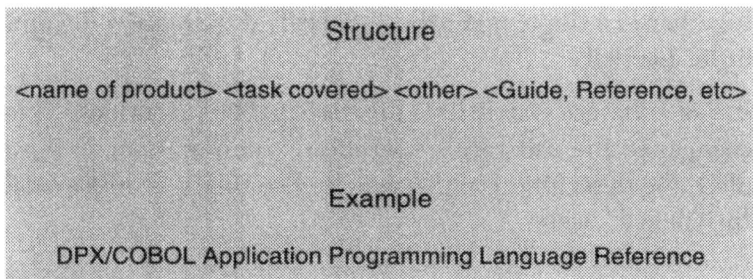

Structure

<name of product> <task covered> <other> <Guide, Reference, etc>

Example

DPX/COBOL Application Programming Language Reference

Figure 7.2   Titling in mainframe manuals.

7.8.1 User's guide

The printed information element, *User's Guide to OS/2 Warp,* has four parts and three appendixes.

The four parts are:

- Part 1—Installing OS/2.

- Part 2—Exploring OS/2.

- Part 3—Using OS/2.

- Part 4—Troubleshooting (Here, you will find an explanation and action for each error message that can appear when users install the product).

The three appendixes are:

- Appendix A. Keyboard and Mouse Use.

- Appendix B. Books for OS/2 Version 3 Application Developers (Here, you will find brief descriptions of a large number of printed manuals that users who develop applications for OS/2 can order).

- Appendix C. Notices.

7.8.2 Online Help

Online Help is available for every desktop object, every menu item, and wherever there is a Help push button.

- *F1 Help:* Provides information about objects, pop-up menu items, entry fields, and push buttons.

- *Help push button:* Describes most of the object pop-up menus and notebook pages.

7.8.3 Online books

During installation of the product, the online information and books shown in Table 7.3 are added to the Information folder.

Table 7.3
Information Elements for IBM OS/2 Warp

Information Element	Description
Master Help Index	Provides help for almost everything the user may want to know about the OS/2
Command Reference	Describes all the commands that can be used at the OS/2 command prompt
Glossary	Defines the terms used in various OS/2 information elements
REXX Information	Helps users get familiar with REXX language and programming concepts
Windows Programs in OS/2	Explains how to use OS/2 and Windows programs together under OS/2
Application Considerations	Explains how to enable certain programs to run under OS/2
Performance Considerations	Provides information about system performance improvement, memory management, communication ports, and so on
Printing in OS/2	Explains everything from installing a printer to solving printer problems
Multimedia	Describes multimedia programs available with OS/2
Trademarks	Describes trademarks mentioned in various online information elements

References

[1] Constantine, Larry L., "Toward Usable Interfaces: Bringing Users and User Perspectives into Design," *American Programmer*, Feb. 1991.

[2] Rubin, Jeffery, *Handbook of Usability Testing: How to Plan, Design, and Conduct Effective Tests*, New York, NY: John Wiley & Sons, Inc., 1994.

[3] Gould, J. D., and Lewis, C., "Designing for Usability: Key Principles and What Designers Think," *Communication of the ACM*, Vol. 28, No. 3, March 1985.

[4] Brockmann, John R., *Writing Better Computer User Documentation: From Paper to Online*, New York, NY: John Wiley & Sons, Inc., 1992.

[5] Watts, Doris R., "Creating an Essential Manual: An Experiment in Prototyping and Task Analysis," *IEEE Transactions on Professional Communication*, Vol. 33, No. 1, March 1990.

[6] *User's Guide to OS/2 Warp*, IBM Corporation, 1994.

8

Designing the Specifications and Prototypes

ow that you have the information architecture, you know which information elements need to be written. However, before you can write, you must design. During the UCID low-level design phase, you will design each information element—keeping in mind the goals you set earlier. First you will specify the content and organization of each information element. This activity is referred to as designing the specifications. Then you will create prototypes or models that represent each element (or a distinct section of it). The rest of the element will have to be written like the prototype. This second activity is referred to as designing prototypes.

8.1 Designing the specifications

Once you have designed the information architecture, you should design and plan for the development of each information element. The specification for an information element is the low-level design of the content and organization of the information element. In addition, specification includes all details another technical writer can use to successfully complete the project without further inputs from the original writer. Specifically, it should cover the following details for each information element: purpose and users, outline, critical design goals, and design and production issues.

8.1.1 Why develop specifications?

Specifications help you better understand how useful the information architecture and the information elements are for the users. With specifications, the managers, reviewers, and users know that you have considered the users' needs.

Specifications reveal content and organization flaws, and allow for early review and significant improvement. You can review headlines for clarity and retrievability. In contrast, opportunities for significant improvements are fewer if you ignore this step and submit for review, say, a 200-page information element. You will also avoid reorganizing late in the project, which can be costly and time-consuming.

Specifications support collaborative writing. The big picture is clear. Everyone knows what each is writing. Specifications also greatly help when new writers join the team. As it explores each information element in detail, specifications allow you to better estimate the effort for information development.

8.1.2 Specification design process

Specifications are the output of the UCID low-level design phase. They can be a part of your UCID plan or, if large, can be a separate document. Specifications are prepared by the technical writing manager or a senior technical writer.

You should plan for the specification effort. Outlining, an activity that is part of designing specifications, can take up a significant part of the specification effort. Moreover, you should allow time for the evaluation, iteration, and improvement of the specifications.

Cover the following details in your specifications.

- *Purpose and users.* You may want to repeat some details from the information architecture. State the purpose or use of the information element. Include a brief statement of how the information element helps meet software usability goals. Also, list the user groups that will use the element.

- *Outline.* This is the main part of the specifications. Read the next section to learn how to develop outlines for your information elements.

- *Critical design goals.* List the critical design goals for each information element. See Chapter 6 on how to design CDGs.

- *Design and production issues.* List all the design techniques and other devices you want to use in order to achieve the information quality goals. For example, describe the retrievability devices you want to provide. Include any special production requirements such as hard separator pages for printed elements. Also include page or screen count and the estimated number of graphic items.

8.1.3 Designing outlines

You should design an outline for every information element. Your outline shows all the pieces of information (topics) that are best used from the information element in question. It also reveals how users will retrieve the information.

Why you need an outline From the users' viewpoint, there are two important benefits. First, the outline helps ensure that all topics they require are included and that all topics not required are avoided. Second, the outline helps ensure that the topics are organized the way they expect. Unbalanced headings show up clearly in an outline and can be easily corrected.

From your organization's viewpoint, there are other advantages. The outline is the framework that guides the writing effort. It provides a quick and easy-to-grasp overview of the entire information element. The outline helps in collaborative writing. It ensures that writers are assigned for all the topics. It lets you see the element as a whole and better integrate the parts. Estimating effort for developing the information element is easy. Reviewers can detect omissions, unnecessary repetitions, and illogical connections. And at this stage, corrections are less time-consuming and less costly.

How to create   A good outline is rarely the result of the creativity or hard work of a single technical writer. For best results, you should gather a group of interdisciplinary people for brainstorming. The brainstorming team should have technical writers, programmers, usability engineers, maybe even marketing specialists. Before starting a brainstorming session, the technical writer should do some homework. Identify an existing software system that is similar to the one you are working on then pick an information element whose content, purpose, and target users are similar to the one you are planning to design. This is a good place to start. In fact, this is one way to find some of the users' information requirements.

At the brainstorming session, list all relevant topics. Invite comments from others. Once you have everyone's agreement on the list of topics, categorize the topics into related groups.

Once you have grouped topics, organize the topics within each group, then organize the groups themselves. The three most important things to remember in organizing information are (1) information should be organized in ways users expect, (2) organization should be apparent to users, and (3) organization should promote retrievability. Some organizing approaches to consider are chronological, order of use, and alphabetical. See Chapter 13 for discussions on organizing approaches.

Pay particular attention to the way you word the topic headings. Information will be hard to find if headings are not written from the users' viewpoint. Place headings of equal importance at the same level. Indent or assign a number or a letter to each heading so that headings of equal importance show a relationship to each other. Use different typography or successive indentation to show subordination. See Figure 8.1.

Under each topic (or heading) include a brief summary of the material to be presented. However, do not provide this if you cannot write a

1. Introducing ABC

Part I. Setting up ABC

2. Defining customer and account details
   2.1 What are customers and accounts
   Description: Make it clear to users what the
                the terms customer and account
                mean in ABC.
   2.2 Maintaining customer details
   Description: After a brief introduction to the
                concept of "customer," list the steps
                required to add a customer. Then
                say what could be done after
                adding customer details.
   2.3 Maintaining account details
   Description: Briefly bring out the relationship
                between customer and account.
                List the steps required to add an
                account. Then say what could be
                done after adding account details.

3. Defining services and billing units
4. Setting up billing periods

Part II. Using ABC

5. Forecasting resource use
6. Preparing a budget
7. Billing customers
8. Producing reports

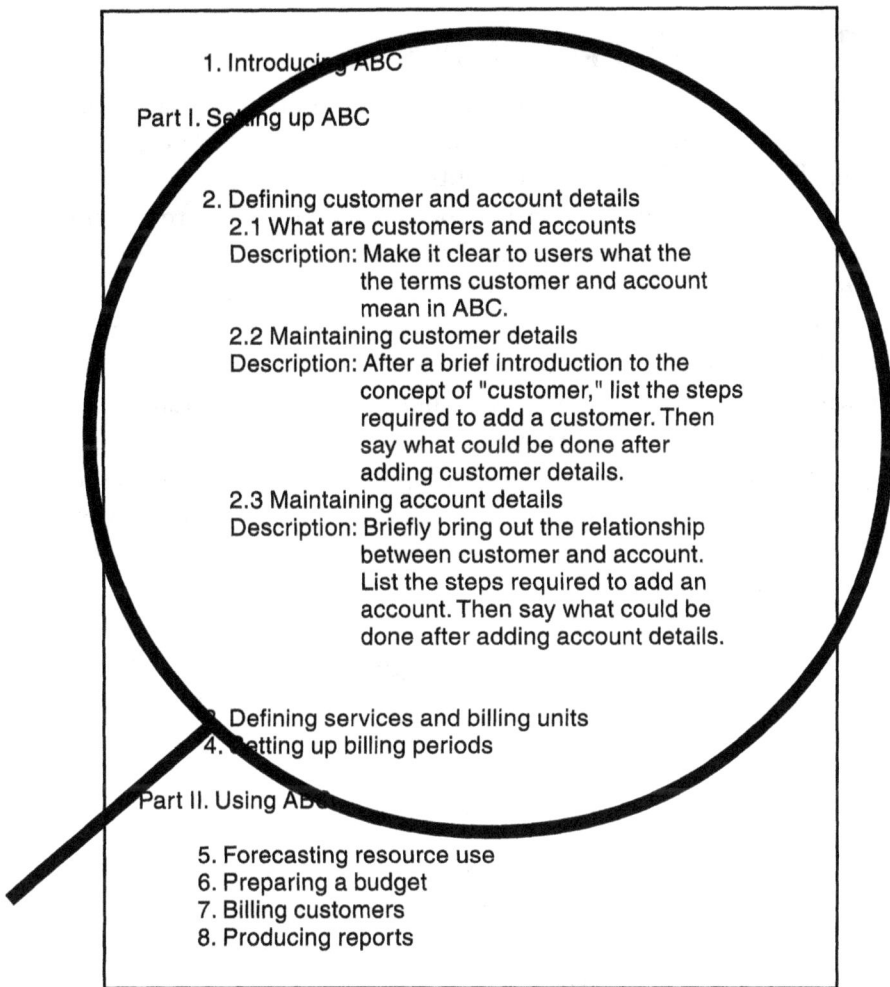

Figure 8.1 A sample outline.

useful summary. For example, "Users will learn how to test information" is not useful if the headline is "How to test information."

A computer-based outliner such as the one you find in sophisticated word processors will greatly help in organizing your outline, allowing you to easily and quickly make changes.

8.2   Designing prototypes

Many industries have used the prototyping approach. Many software houses do user interface prototyping. Some have extended prototyping to the design of information elements.

In this book, the word prototype is used to mean a representative model of an information element. An example for a prototype could be a chapter of a user's guide. The prototype will have all the items you plan to cover, complete with final draft-like formatting and figures. The prototype could also be a representative model of a distinct portion of an information element. Prototypes of online support elements should preferably be working models, though of reduced functionality. These prototypes are used to build the rest of the chapters or topics they represent.

The prototype should be iteratively evolved through tests, reviews, and editing. Working models of the information element are incrementally created via design, evaluation, and iteration. Early attention is on the need for (and suitability of) the information element. Subsequent iterations include attention to the critical design goals defined for the element. To get the best results, the prototyping process should remain open to major changes at any stage.

8.2.1   Why design prototypes?

The primary use of a prototype is as a model. It helps bring in consistency across chapters or sections that you develop based on the prototype.

The prototype helps investigate a new design idea such as a new navigation technique, a new organization approach, or a new page (or screen) design. It helps you try out many designs and pick the one that best meets users' needs. You will avoid designs that do not contribute to software usability. Early prototyping saves time by reducing the need for costly late changes.

The prototype helps reviewers and users provide important feedback. Reviewers generally have difficulty responding to just an outline, especially an outline for an online information element. Reviewers, especially users, may review and agree to specifications and yet dislike the final result. A fully conceived prototype portion of an information element will fill the gap between specifications and a draft of an information element.

Once you have the prototypes, you can refine your effort estimates. For example, based on the time it took you to complete the prototype for a user's guide chapter, you can arrive at a more accurate estimate of hours per page.

8.2.2 Prototype design process

Before you begin prototyping, you must have the information architecture and the specifications. Specifications must be available for all the information elements you want to prototype.

Plan for prototyping, evaluation, rework, and iteration. Document the prototypes in your UCID plan or, if large, in a separate document. Indicate what the model represents and where in the outline it fits in. Prototypes are primarily prepared by senior technical writers in consultation with the technical writing manager, programmers, and usability engineers.

Here are the things you should do:

- *Decide what to prototype*. Identify a distinct portion of an information element to prototype.

- *Design and develop prototypes*. Fully implement the design in terms of organization, content, and interaction (if online element). If it is a prototype for an online information element, you may need programming support.

- *Evaluate*. Reviewers must include actual or representative users. The information must also be edited by professional editors to conform to the standards you will follow in the project. If you are evaluating alternative designs or checking out new approaches, you will want to organize a formal test. If your model is a complete chapter and the task it covers is the task that your project team has decided to test, you can have it participate in the integration evaluation. (Typically, however, you will need at least a few chapters for integration evaluation.)

- *Iterate*. Evolve the prototype through evaluation and redesign. Be prepared to throw away unsuccessful designs and start over.

8.2.3   Designing and evaluating prototypes

In what kinds of projects should you attempt prototyping? A prototype for every information element proposed in the information architecture is a must for all UCID projects. In fact, you might even have two or more for a single information element. An example situation is when you have an element titled *User's Guide and Reference*. Here, you will need two models, one each of the guide and reference types of information.

You may not need too many iterations if you are evaluating standard or familiar ideas or new versions of existing software. Iterations that are the most beneficial are the ones where you are considering a major design change.

Which part of each information element should you prototype? The answer is, those with the following characteristics:

- Representative of the entire information element or a distinct type of presentation within an information element. Create a prototype for different types of tasks (e.g., one for data entry tasks and one for report generation tasks). Or, create prototypes for pieces of information aimed at clearly different user profiles. See Figure 8.2.

- Large enough to be evaluated and adhered to.

- Sufficient source material available to complete all the items such as text and figures.

You should evaluate prototypes via CDG evaluation. Consider evaluating many alternative prototypes simultaneously.

2. Defining customers and accounts

This chapter tells you what customer and account mean in ABC. It also tells you how to add, modify, and delete the details of customers and accounts. Clerical staff at the computer center can perform these tasks as soon ABC has been customized to your requirements as described in ABC Installation and Customization Guide.

2.1 What are customers and accounts

A customer is an external client or in-house department to whom bills can be sent or costs allocated for the use of your computer resources. An account in ABC is a person from the customer organization who actually uses the resources.

2.2 Maintaining customer details

To charge the customer for resources used, you should identify the customers, classify them as either client or department, and add them to ABC.

To add customer details:
1. Select Customers from the ABC main menu.
2. Select Add from the Edit pulldown menu.
3. Type the required and other details.
4. Press Enter.

Now you can modify customer details or delete customers using the Edit pulldown menu.

2.3 Maintaining account details

An account in ABC is the person who actually uses your resources. You can add one or more accounts for every customer you've earlier defined.

To add an account:
1. Select Customers from the ABC main menu.
2. Select a customer who should be charged for resource use by the account.
3. Select Accounts from the Associate pulldown menu.
4. Select Add from the Edit pulldown.
5. Type the required and other details.
6. Press Enter.

Now you can also modify account details or delete acounts using the Associate and Edit pulldown menus.

Figure 8.2   A sample prototype.

9

Designing Labels

L ABELS ARE EVERYWHERE—on doors, stoves, dashboards, and book shelves. We attach labels to items so that people can identify those items. When we cannot remember the label, we can still locate something because we remember that it was, say, put in a box of a certain shape, size, and color. Such richness of detail found in the physical world, however, is not yet available in the software world. Software labeling, therefore, is a critical and difficult part of user-centered design.

This chapter is about labeling in the software world. Technical writers should read this chapter before they attempt labeling.

9.1   Labels and software usability

In a software system, labels identify user objects (menu options, etc.) and commands. Through text, icons, or text-icon combinations, labels communicate meaning, content, action, or state. Labels are *interaction information* because they are essential for user-software interaction to go on successfully.

In the software world, inadequate labels are numerous. Some labels are vague, leaving users uncertain about what to do next, while others are misleading, promising one thing, but bringing up something else. This could be because of inconsistent labeling. For example, the user is confused if the label for a dialog box is different from the label of the pull-down option he or she selected.

In typical software engineering projects, labels are first created by analysts and programmers. These terms are used in the early requirements specification stages and become part of the software system. Some of these labels are meaningful to users while many are only good in communicating with other technical members of the project team.

Labels are very important because of the intangibility of software. The quality of the label determines the usability of the user object or command that the label identifies. Effective labels can make a software system easy to navigate. They lead users through the steps required to complete tasks. Moreover, they encourage the desired capability of "user exploration" if the software is designed with emergency exits and facilities such as undo and backtrack. Good labeling helps avoid errors and improves user performance. It can also mean less support information. And that is good news for most users and software houses as well! So, you should spend extra time to make labels effective. Your extra effort could save many people weeks of writing explanatory support information. More important, it could save many user hours spent looking for and attempting to understand the labels. Another advantage of effective labels is the improved retrievability of support information via an index. In an index, appropriate labels are the only cue users have to quickly finding the required information.

9.2   Label definition process

User/task analysis is over and you have just entered the UCID high-level design phase. Product prototype design is about to begin. And prototype designers are going to need labels. Although you are now making formal efforts to define labels, even during task analysis you may have already decided on some of the task-related labels.

Good labeling demands planning, creativity, even some research. Technical writers must work closely with analysts, programmers, usability engineers, and users to design labels and manage the label definition process. If the software system supports languages other than English, you should plan for translation. The label definition activities are as follows:

1. *Design labels*. Design textual and icon labels from the users' viewpoint. For icon labels, get help from a professional graphic designer. You should, however, brief the graphic designer and also review the icons.

2. *Define glossary*. A dictionary-like formal glossary is required to make sure everyone in the project knows *what* labels should be used, understands the meanings of labels, and uses them consistently. Remember that labels include task names, which are also used in support information elements. Once the glossary is ready, distribute copies to all the project team members. Also, attach the glossary as part of the UCID plan.

3. *Control terminology*. This is an ongoing activity. As the software system evolves, new labels may be required. Every time a need is felt for a new label, ask yourself if users really need a new label or if an existing term would do.

4. *Evaluate*. Set up a team to evaluate and approve labels. This team should include analysts/programmers, usability engineers, users, and editors. At the end of *integration evaluation*, you will probably have some feedback on the effectiveness of labels. In addition, you should organize *CDG evaluations* with users to evaluate and improve the effectiveness of labels.

9.3 Labeling

... a label is about what's in the box, but the question is what's in a label?

Lon Barfield asks this question in a thought-provoking and interesting article [1]. While moving to a new house, Barfield attempts three ways to label the boxes: first, based on the contents (books, cutlery, etc.); second, based on the area that the contents came from (mantelpiece, kitchen drawer, etc.); and third, based on the area where they should go. How would you label in the software world? What goes into a label? Should a label be text or icon, or a text-icon combination? How will you decide? What are the constraints and considerations? The following sections attempt to answer these questions.

9.3.1   Labeling with text

User object labels   User objects can be windows, menu options, fields, columns/groups, buttons, and combination controls. Communicating the usefulness of a user object in just a word or a short phrase is a challenge (see Figure 9.1). Task labels are perhaps the most important type of labels because they identify tasks users perform with the software system.

People who design the interaction components may provide keyboard shortcuts for menu options. For that, a letter selection technique is used. It requires a unique letter (preferably the first one) for each option. A problem arises when two options have the same first letter, such as with Maintain and Modify. Avoid the solution of renaming one of the options if the labels are otherwise effective. If you must rename the term Modify, try alternative terms like Adjust, Alter, Change, Edit, Replace, Revise, or Update. You will find such synonyms in your thesaurus. Always keep in mind that you need labels users will understand.

Command labels   Command is a label that is typed in by users. Most user objects are visible, but commands are not. Choosing a menu option is usually easier than remembering and typing command syntax. Therefore, command labels must be more carefully designed to support user recall. Command names are often abbreviated forms of English words or phrases. Examples are LCD for "local change directory" and CHOWN for "change owner." Whereas such abbreviations are very efficient for expert users, they can be difficult for novices—especially if there is a large number of commands.

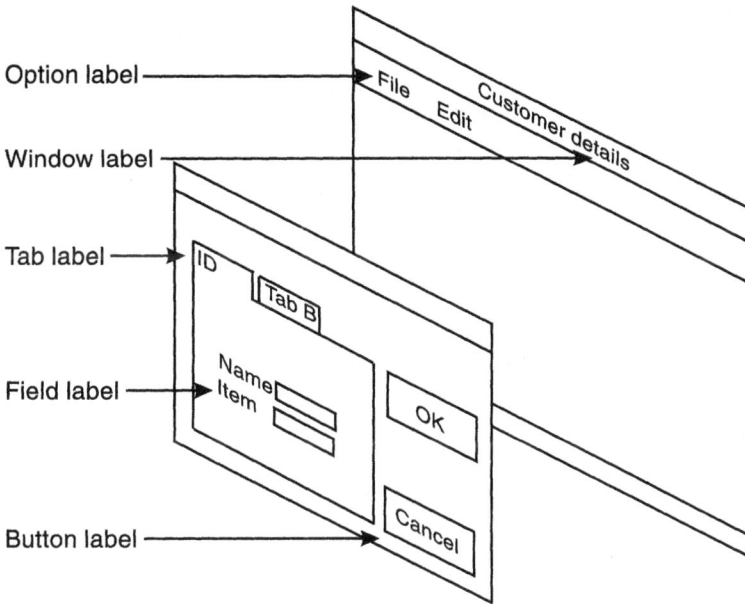

Figure 9.1 Textual labels for user objects.

Label control Do not create a new term if the concept can be expressed adequately and accurately by a simple combination of English words. A new label may be required when the idea must be expressed frequently in writing and when the number of words needed to express that idea is too large. Before adding a new label, the evaluation team set up for this purpose should consider that term. The team should ask questions such as the following:

- Do we really need a new label?
- Is the label unique, without a definition in a standard dictionary?
- Can it be used in other products if it represents a general concept?
- Is it understandable?
- Is it acceptable to international users?

If the answer is yes to these and other project-specific questions you may have, then go ahead and add the new label.

Defining a glossary   After a required new label is designed, you need to provide an accurate definition for that term from the users' viewpoint so that it can be understood properly and applied consistently. Here are some guidelines:

- List glossary entries in alphabetical order.

- Use generic terms rather than coined terms.

- Use English words only as defined in standard dictionaries.

- Avoid distorting terms to obtain acronyms.

- Do not write definitions that conflict with the meanings of established terms.

- Do not define a word by using the word being defined or a variation of it. For example, do not define "allocation" as "The process by which resources are allocated to a CPU." This definition does not help the user because it defines an unknown word by using a variation of the same unknown word.

- Do not define a noun as a verb or vice versa. For example, the following two definitions would be incorrect if they were interchanged: "Justify: To adjust the printing positions of characters...," and "Justification: The adjustment of...."

9.3.2   Labeling with icons

In graphical user interfaces, icons are used to indicate the status of the software system, to warn users about an undesirable event, as well as to label user objects. When designing icon labels, design them as a set, considering their relationship to each other and to the users' tasks. See Figure 9.2 for various icons used as labels:

While talking about computer hardware labeling, [2] argues that text in a label is only essential if the following three conditions are true:

1. No well-understood icon is available;

2. The function is abstract (e.g., "reset");

3. The consequences of misinterpretation are severe (e.g., loss of data).

Figure 9.2 Icon labels.

This appears to indicate that you should first attempt an icon label and then consider text if the three conditions are true. In the software world, however, it is a good idea to have text either alone or in combination with an icon, and make the display of either optional. The Netscape Navigator browser allows users to display the main toolbar as icons, text, or combination. In an icon test, users were given the name of an icon and a short description of what it was supposed to do, then asked to point to the icon that best matched the description. The group of icons that included text as part of the icon got consistently high scores.

Labeling with text, of course, has some disadvantages, especially for international products. Words from some languages will not be able to fit into the space provided for, say, pull-down options. Textual labels would need to be translated and printed in multiple languages. Producing multiple-language labels could be expensive and difficult to implement.

9.4   Critical design goals

If we consider users' requirements, common label design problems, and label characteristics that improve software usability, we can shortlist five critical design goals for labels. The goals are listed below:

1. Clear;

2. Distinctive;

3. Short;

4. Consistent;

5. Pleasing.

9.4.1   CDG 1: clear

While designing labels, one key software usability principle you should follow is "Speak the users' language." Labels should be based on the users' language, not on the system's. Extend your concern for the users' "language" to icons, too.

Speaking the users' language does not always imply limiting terminology to a few commonly used words. Instead, if a user group has its own specialized terminology for its application domain, the software system should use those specialized terms rather than the less precise, everyday language.

Dieli and others [2] say usability tests proved that Microsoft Excel users could not easily match functions such as Consolidate, Scenario Manager, or Parse with tasks they wanted to perform. If users wanted to combine several ranges, they can neither map that task to the Consolidate function nor find information in Help easily without knowing the name of the function. You must, therefore, know the users' vocabulary to describe tasks. For example, users who want to consolidate several ranges in a spreadsheet may say "combine" or "totaling ranges" for that task. You should elicit these terms from them. Use of such user vocabularies should significantly increase users' ability to match labels to tasks. Try beginning with active verbs, as in "Maintain customer details," because the users are performing tasks. Use the same verbs consistently, even in other information elements.

Display labels in a way that reinforces their purpose. Suppose you have menu options about the alignment of text. The options could appear in this order: Left, Center, Right [3]. The order reflects what the options do. The Style list on the Formatting toolbar of Word 97, called the Style Preview, shows the actual formats associated with styles. Here, the label's purpose is visually reinforced. One disadvantage of this technique is that if the list is too long, it may require a lot of scrolling. See Figure 9.3.

Make the labels specific. Rather than a general "Open File," consider a more specific "Open patient details." Does the label "Transfer" mean to go to another application, swap files, or move data to another application? In the last case, "Export data" might be clearer.

For clarity, consider using complete words or phrases wherever possible. For example, "Delete" rather than "Del."

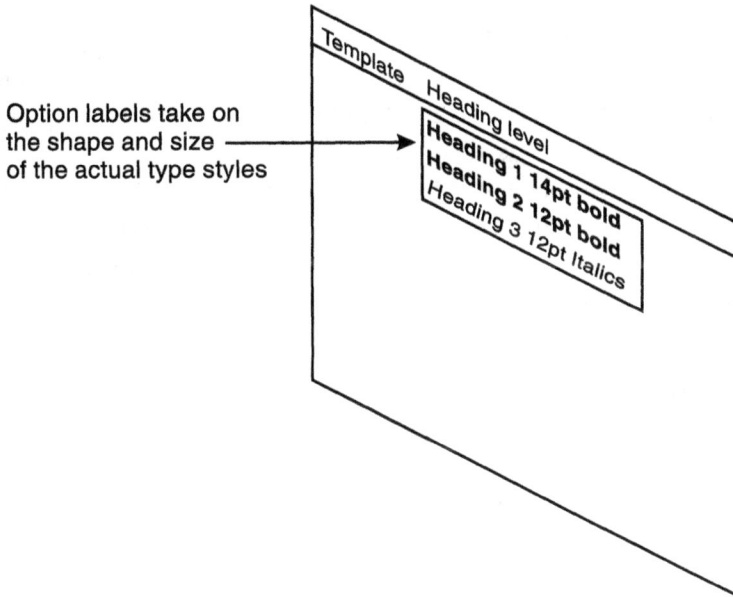

Option labels take on
the shape and size
of the actual type styles

Template   Heading level
Heading 1 14pt bold
Heading 2 12pt bold
Heading 3 12pt Italics

Figure 9.3   Reinforcing the label's purpose.

Consider international use. Words and abbreviations can carry unintended meanings outside the English language. A spreadsheet with the abbreviation SS in the title would have negative connotations in Germany because of the dreaded storm troopers in World War II [4]. Do not use words like "Abort" (abortion being an emotionally charged issue).

I have read a case where the *International Organization for Standardization* (ISO) required that icons be correctly interpreted by at least 66% of the users in a test in order for the icon to be considered for adoption as an international standard. To test the intuitiveness of the individual icons, show them to the users, asking the users to point out the meaning. Since icons are normally seen in groups, test the understandability of groups of icons by showing groups of icons.

9.4.2   CDG 2: distinctive

Distinctiveness facilitates memory and retrieval. Labels should be *distinctive* so that users will not confuse one with another. Avoid subtle differences, such as between "View" and "Show." Users may try to use them interchangeably. Pick names that use different letter patterns. Do not use a series of options or commands that all begin with the same letter, such as Delete, Demote, and Duplicate. Such a series is easily confused. "Customize ..." is distinctive because of the trailing dots that indicate that a dialog box would appear when the label is selected. Distinctiveness is especially critical in command labels. Incorrect command entry due to lack of distinctiveness may produce valid, unintended commands.

9.4.3   CDG 3: short

Labels have to be fit into very little space, especially in pull-down menus. Therefore, they need to be short. You can keep labels short by avoiding grammatically complete sentences. For example, "Export file" is acceptable and preferred to "Export the file." Keep labels short, but clear. That is a challenging task.

9.4.4   CDG 4: consistent

Consistency is getting a dialog box labeled "Convert Text to Table" when you select the option labeled "Convert Text to Table." Consistency is also parallel structure of a group of options. For example, the labels Cut, Copy, and Paste are consistent because they are all verbs.

Consider if users would expect labeling consistency across products.

9.4.5   CDG 5: pleasing

To design icons, use a professional graphic designer. Icons should be created to display well on the screen. You should also consider international users while designing icons.

9.5   Integration considerations

If you only provide icon labels, you may want to provide a word or two of description in a small pop-up windows such as in Figure 9.4. If users will type commands, you may want to focus on online Help as a key information element.

Figure 9.4 Icon label and supporting Help.

You should consider providing online Help for user objects, especially for fields where users need to type data. For example, you may want to provide Help such as shown in Figure 9.5 for the Name field.

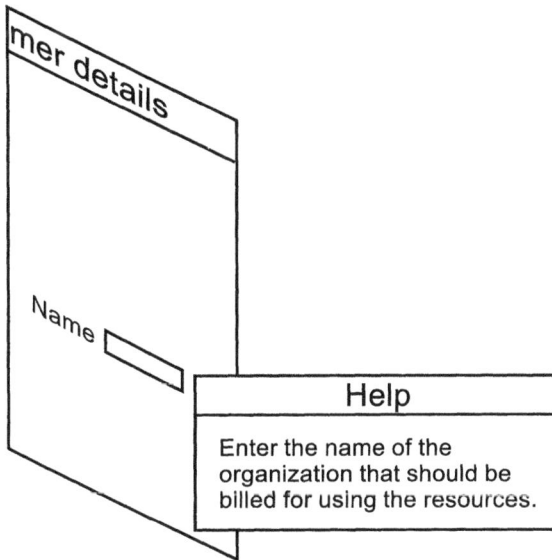

Figure 9.5 Online Help for entry field.

References

[1] Barfield, Lon, "Sticky Labels," *SIGCHI Bulletin*, Vol. 28, No. 2, April 1996.

[2] Wiklund, Michael E., *Usability in Practice: How Companies Develop User-Friendly Products*, Cambridge, MA: AP Professional, 1994.

[3] Powell, James E., *Designing User Interfaces*, San Marcos, CA: Slawson Communications, Inc, 1991

[4] Ferandes, Tony, *Global Interface Design: A Guide to Designing International User Interfaces*, Chestnut Hill, MA: AP Professional, 1995.

10

Designing Messages

Effective communication takes place when it is two-way. In user-software interactions, two-way communication is critical. Providing feedback to users is more than just a basic requirement for civilized manners. As it can draw various undesirable reactions from users—panic, confusion, frustration—system feedback significantly impacts the usability and success of a software system. Feedback is commonly called a message. In this chapter, I have used the two terms interchangeably.

A message is any software-initiated display of information in response to a user action or change in system status. Contrast this with user-initiated information display such as online Help, which is described in Chapter 11.

This chapter is about designing messages. Technical writers and programmers who coauthor messages should read this chapter.

10.1 Feedback and software usability

Knowledge of results, or feedback, is a necessary learning ingredient. It shapes human performance and instills confidence [1].

When the user requests something, the software cannot sit quiet—irrespective of whether it can and will attempt the user-requested job. When users initiate a job, for example when they click the Check button in a virus check program, the system should immediately give feedback to confirm that the job has started (or failed). This feedback could be a message like "Started checking...." While the job is going on, the system should give adequate feedback about the progress and status. Finally, when finished, it should inform the user about the completion of the job, indicating whether or not the scan was successful.

People who developed the software may think that all this feedback is superfluous and unimportant. That is probably because they are familiar with how the system works. But users neither have that knowledge nor have the interest or need to know such details.

Not only is immediate feedback required, it should also quickly help the user understand and correct any potential or existing problem.

I have occasionally used a mainframe operating system. Recently, I tried to *allocate* (create) a new *data set* (file) by specifying various details including the data set name. When I pressed Enter after typing the details, an error message appeared, as shown in Figure 10.1.

Of course, I did not understand it. I called a programmer colleague to my terminal. He wasted many minutes trying to find *what* the error was.

The mystery became more intriguing. Meanwhile, I had a whole bunch of curious system software experts around my computer terminal.

```
File ISP15535 not allocated, system or installation error.
Text unit x'0002' contains invalid parameters.
```

Figure 10.1 An unnecessarily complicated message.

Finally it struck one of them that the problem was actually ridiculously simple: the data set name I entered did not follow the system's file-naming conventions! The name started with a number instead of a letter. How come they did not simply write, "Data set name cannot start with a number" or something like that? To this day, neither I nor my helpful system software colleagues have a clue.

That message is feedback, alright. But it is feedback that does not quickly help users understand and appropriately respond.

Unhelpful feedback or no feedback can create undesirable situations for users, including loss of their data or piling up of a greater number of perhaps more catastrophic errors.

Typically in the real world, the level of attention given to quality varies depending on whether the software is developed for a specific client, for in-house use, or for off-the-shelf sale. For example, if the software is for a specific client, the software house may not give the required level of importance to the quality of printed manuals. The message set is one information component where companies cannot afford such compromise. The quality of messages has to be good irrespective of the purpose of the software.

Feedback significantly affects user performance and satisfaction with the system. Good feedback reduces the number of problems users have, or at least makes it easy for them to correct potential or existing problems. Consequently, good feedback also reduces the number of user support calls and the cost of user support. Therefore, you will find feedback listed in every set of guidelines (or heuristics) on good user interface design.

10.2   Why users have trouble with messages

Messages often bring down the usability of software for many reasons, ranging from just inelegant wording to seriously incorrect statements.

Perhaps the most common problem is that they are designed from the system (or programmer's) point of view, rather than from the user's point of view. Users describe such messages as unclear, misleading, meaningless, and too technical. The result is that they cannot quickly continue with the overall task they were attempting. The message "End billing period must be greater than start billing period" is an example of system-oriented thinking. If it were written from the user's point of view, it would have been "End billing period must be *later* than start billing period." In the real world, there are numerous system-oriented messages that are more harmful than this example.

Some messages are cryptic, sometimes thinning down to meaningless codes such as "SYS2030" and "TRAP0002." With such messages, users are forced to search for more helpful information elsewhere.

Misleading messages can be as bad as messages that give wrong information. Some messages can mislead simply because they are incorrectly worded or contain an incorrect tense.

Unhelpful messages such as the classic "Invalid entry" and "Illegal action" continue to be written. If you are thinking of writing "Error encountered in date format" you should probably consider turning that into "Date must be in the MMDDYY format."

In a home design software, the Shift Amount field requires a number to be entered. If you enter a letter or any nonnumeric character, the resulting error message says: "Shift Amount is Zero." Perhaps thought was not given to the various possible reasons for the error—by mistake, users can type any character on their keyboard, not just a zero.

Then there are messages that are offensive or sound inappropriate in other ways. Frightening messages such as "Fatal error" cause novice users to believe that they made a mistake due to their own lack of skills—when in fact the error probably occurred because of poor interaction design. Figure 10.2 shows a message that appeared on my computer.

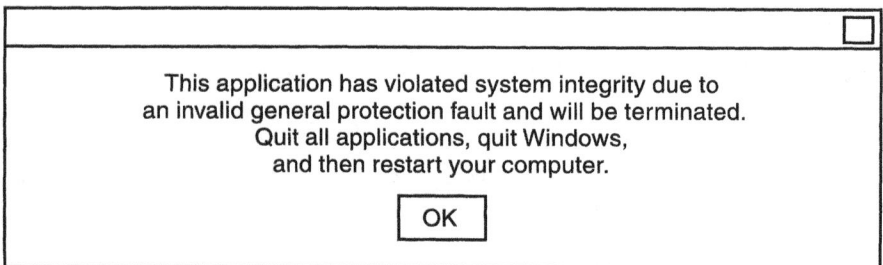

This application has violated system integrity due to
an invalid general protection fault and will be terminated.
Quit all applications, quit Windows,
and then restart your computer.

OK

Figure 10.2   An unnecessarily frightening message.

After clicking on the OK button, I had to only double-click on the Application icon to get right back to the application I was using. None of the user actions mentioned in the message was required!

Messages that are poorly worded can often be corrected by a good technical editor, even at a fairly late stage of the project. On the other hand, messages that are system-oriented or out of context can be hard to fix.

Feedback should only be provided where necessary, too much can be unnecessary or even harmful. For example, trivial informative messages that frequently attract attention quickly get users immune to them. Users then begin to ignore the really important messages.

10.3 Message design process

Message design is traditionally a programmer's job. In a study described by Duffy and others [2], only 22% of experienced technical writers indicated that they were responsible for writing messages. However, message design is an area of user-software interaction that clearly benefits from good technical communication techniques. Therefore, although programmers have traditionally written messages, technical writers and editors have often been called in at a later stage to improve the quality of messages. In the UCID project, technical writers will collaborate with programmers and coauthor messages.

You will write messages during the UCID development phase, which is alongside the development and testing phase of your software project. You will also need to write a few messages during the user interface prototyping stages.

The activities related to message design are as follows:

1. *Identify message categories.* You are most likely to have the three standard categories: informative messages, warning messages, and error messages. At the time of designing the information architecture, you will have also decided what types of information will go into each category. For example, will error messages only mention the error or only give the corrective action, or provide both?

2. *Design message display*. Design how messages will be displayed on the users' screen. Do this before you begin the development of a substantial number of messages.

3. *Write messages*. It is a good idea to plan for reuse before you begin writing a large number of messages. In a project I worked in, "Invalid entry" was reused in about 20% of the 800 messages! That is an idea considered terrific by the development team and would have probably been loved by the product maintenance team. Weigh the consistency and productivity benefits of message reuse against context-specific messages, which are more helpful for users.

4. *Evaluate messages*. As part of CDG evaluation, you should test messages. In Microsoft Word 4.0 for DOS, if you enter Word, type some text, and select Quit, the message, "Enter Y to save, N to lose edits, or Esc to cancel" appears. Pressing Esc cancels your selection of the Quit option. Typing N quits Word without saving your text. And typing Y—rather than asking for a file name to save the text—displays the message, "Not a valid file." Such flaws can only be uncovered by testing, not by reviewing a list of messages printed on paper. Test if the right message appears at the right place. "Create" errors and check if the resulting messages are appropriate for the error situations.

10.4   Common categories of messages

The three basic message categories are as follows:

- Informative messages (to inform);
- Warning messages (to warn);
- Error messages (to indicate error).

Message classification is required because you should use a different display technique for each category so that the seriousness of the message is immediately perceived by users. More important, the type of information that goes into each category will also be different.

Because the display and content of each category is different, you can also think of the three message categories as three different information elements.

10.4.1 Informative messages

Inform users about what *has* happened. Figure 10.3 shows an example of this.

In case of a continuing job, provide adequate feedback about what *is* happening—the progress and status of the job. An example of this is "Updating general ledger ...."

The user interface should let users know when to expect delays, why, and for how long. In *graphical user interfaces* (GUIs), the hourglass is a good way to graphically say "Please wait." On the IBM mainframe, an "X SYSTEM" message appears. Such feedback has been found to work for short intervals of up to a few seconds. For longer delays, it is important to give users the maximum wait time or even a continuous indication of the amount of the job completed against the total. Otherwise, users may start pressing all kinds of keys because they think the system has crashed. This can have catastrophic consequences if the software responds to all those key presses.

10.4.2 Warning messages

Warn users that something undesired might happen if left unattended. No immediate response may, however, be required for them to proceed with the current task. It is often said that a warning is displaying the error message before the error occurs. A warning message can therefore be thought of as a helpful error-preventing technique. Figure 10.4 shows an example of this.

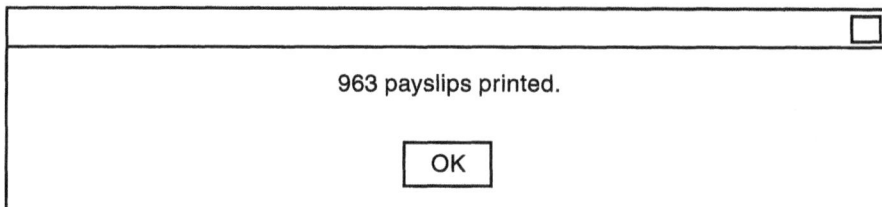

```
+-----------------------------------------------+ [ ]
|                                               |
|              963 payslips printed.            |
|                                               |
|                  +--------+                   |
|                  |   OK   |                   |
|                  +--------+                   |
+-----------------------------------------------+
```

Figure 10.3   An informative message.

```
┌────────────────────────────────────────────────────────┐ ┌──┐
│                                                          └─┤  │
│        Number of columns specified exceeds maximum of 10 allowed. │
│                                                             │
│                        ┌──────────┐                         │
│                        │    OK    │                         │
│                        └──────────┘                         │
└────────────────────────────────────────────────────────────┘
```

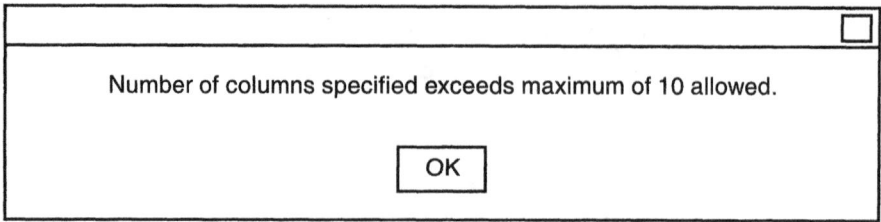

Figure 10.4 A warning message.

As you can see from the example, the user can continue with the task—the system goes ahead, ignoring any columns over 10. However, if the user still prefers to have over 10 columns, he or she can stop and go for, say, a table format, which allows over 10 columns.

Sometimes it may be difficult to decide whether a message is an informative message or a warning message. For example, the message "Customer details are not updated" may at first sound like an informative message, but could be a warning that computer-generated details do not contain the latest details about a customer.

The MS DOS message "<number of files> file(s) copied" is an informative message when the number of files is one or more. It is a warning message when the number is zero!

When a warning situation arises, users will probably need answers to the following questions:

- What can go wrong?

- What can be done to avoid that?

- What is the system's current action?

- When or under what circumstance can I "correct"?

- Who can "correct" it (end user, system programmer, or system administrator)?

- How are corrections made or where can such information be found?

It is not always possible or even necessary to answer all these questions in one message. Depending on the error situation and user skill levels, just addressing the "What is the system's current action?" question may be helpful.

Warning messages are important. Consider requiring the user to press the OK button (or use some similar technique) so that you are sure that the user has seen and acknowledged it.

Use warning messages with judgment. Nielsen [3] cautions about an overuse of warning messages because the user's answers can soon become automatic.

10.4.3   Error messages

A message such as "Customer ID is not entered" indicates an error—the user did not enter details in a field where entry is mandatory. It also requires an action. To continue, the user must make an entry in the Customer ID field. Use such an error message when you want to tell the user that something has gone wrong.

The key to good error message design is to first realize that users are in the midst of a task and that they need information on how to handle the current error situation so that they can continue with the overall task they were attempting.

Sometimes you may find it difficult to decide whether a message is a warning message or an error message. When in doubt, go back to check the user's task situation. The message, "The specified account does not exist" may sound to the user like, "The specified account does not exist, therefore enter an account that exists," in which case it is an error message. Look at the task situation and you may find that it is a warning. The job will now go on, but the user may later find unexpected results because a particular account was not included.

When an error situation arises, users will probably need answers to the following questions:

- What went wrong?

- Why did it go wrong (all possible causes)?

- What is the system's current action?

- What can or should be done to correct it (this may be different for each possible cause)?

- Who can correct (the end user, system programmer, or system administrator)?

- How are corrections made or where can such information be found?

Decide which of these questions you will address in the system-initiated error message and which ones you will answer elsewhere. Remember that it is not always necessary or even possible to answer all these questions. For example, the Microsoft Money message in Figure 10.5 addresses only the first two questions and the fourth, yet is helpful.

10.4.4 Feedback through batch messages

In large computers, users can perform tasks in *batch*. The time-consuming job of processing a large volume of data can go ahead without preventing users from performing other online tasks. Consequently, these computers log error messages in one place, from where users can later retrieve them. Though not timely, this is feedback, too. Therefore, these messages should also be carefully designed.

Just like for online messages, users need more information to understand and respond to batch messages. You can provide such information in a printed manual.

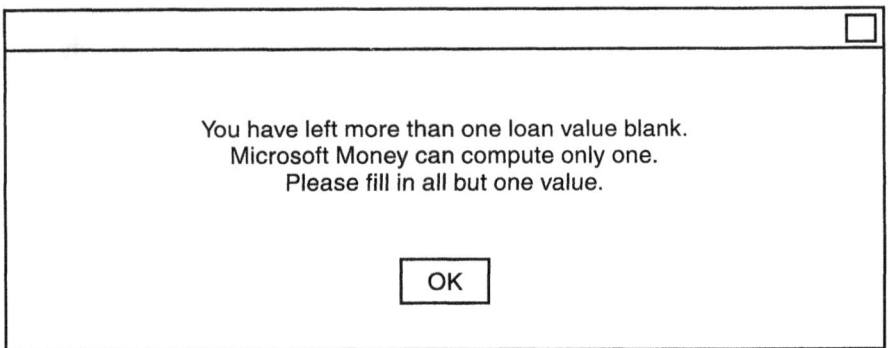

> You have left more than one loan value blank.
> Microsoft Money can compute only one.
> Please fill in all but one value.
>
> OK

Figure 10.5 Helpful error message.

Based on the hardware and software platforms in which the software works, the maximum length possible for batch messages may be even shorter than that for online messages. Because you do not have space to write complete messages, the description you provide in a printed manual should be carefully written. Figure 10.6 shows an example.

Note that this format addresses all the relevant items from the checklist used in designing error messages.

10.5   Interaction design for error

It is often said that the perfect error message is no error and no message. But errors do occur for many reasons, ranging from poor interaction design to user slips and mistakes. It is probably impossible to design sophisticated systems in which errors can never happen. However, here are three things you can still do: (1) minimize the incidence of errors, (2) provide helpful messages for errors that are possible, and (3) make it easier to correct the errors.

Message:

JCL not submitted.

> Explanation:
>> The JCL is tailored to run the batch job, but could not be submitted for execution because of authorization failure.
>
> User response:
>> Submit the JCL after changing the accounting information.
>
> System programmer response:
>> Authorize the user to submit the job.

Figure 10.6   Description for a batch message.

10.5.1 Minimize the incidence of errors

Some techniques that can help minimize errors are as follows:

- Speaking the user's language by using terminology users can understand;

- Providing tools such as those that do the clerical work (e.g., handling typing errors) thereby avoiding human error;

- Representing files as icons or menu options so that users cannot, for example, "select" a nonexisting file;

- Providing a warning message to confirm if a job was intentionally requested, so that the possible catastrophe of an unintended user action could be eliminated;

- Minimizing modes. The system is in a mode if users are required to cancel or complete what they are currently doing before they can do something else. Modes are not completely avoidable, but can be kept to a minimum through careful design. It is interesting to note that providing good feedback is one way of avoiding errors that result from the use of modes.

Every design solution has its own set of pros and cons. Therefore, choose carefully.

10.5.2 Provide helpful error messages

What is a helpful error message? Briefly, it is a message that immediately and pleasantly provides feedback clear enough for users to understand and effectively respond to. A more detailed discussion of what makes a helpful message appears later in this chapter.

10.5.3 Make it easier to correct errors

The goal is to make it easy and pleasant for users to complete the overall task they were attempting. Toward that end, interaction with a system should be designed as a cooperative rather than a finger-pointing activity. For example, undo is a quick error-correcting technique. Users can revert to an earlier state if they are not happy with their (or the system's) action.

10.6   Critical design goals

Before going into the CDGs for messages, let us look at a use model for messages. Users should be able to quickly accomplish the things listed in Figure 10.7.

Based on this model and on common message problems we have seen earlier, and considering our overall objective of improving software usability, we can draw up the eight design goals given below. The goals set assumes accuracy of messages and therefore does not list that item.

1. Complete;
2. Conceived from the user's viewpoint;
3. Context-specific;
4. Clear;
5. Pleasant;
6. Attention-getting;
7. Readable;
8. Removable.

These are desirable characteristics of feedback. Whereas the first five are related to message content, the last three are about message display. Consider the design for feedback along with the design for the rest of the

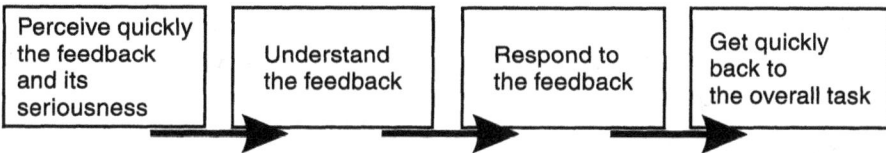

Figure 10.7   Message use model.

user interface, because consistency is important. Display design depends on the seriousness of the feedback. It also depends on the software tools you use and the standards (such as those for windowing systems) followed in your project.

10.6.1   CDG 1: complete

To know the types of information users may need, see the warning and error checklists on pages 140 and 141. Also, read the last section of this chapter to learn how to integrate system-initiated messages with other information elements. The actual content of a message will depend on various factors, which are listed in the following paragraphs.

Factors that influence message content   The following factors influence what information actually goes into each system-initiated message:

- Feedback context;

- User skill levels;

- Actual information requirements;

- Display size restrictions;

- Translation requirements;

- Standards and tools used.

It may be impossible to give all the required information in a single system-initiated message. However, by thoughtfully designing the information architecture, what cannot (or need not) be accommodated in the message can be provided in Help or in a printed manual.

Varying the content to meet user needs   If you have a thorough understanding of the characteristics of representative users, you could try one of these techniques:

- Display various levels of information progressively. Display of information two levels upwards is typically user-initiated. Such information is called online support information and is described in Chapter 11.

- Display a different "text" of the message when users make the same mistake the second time (in the same session).

- Let the user select the length (content) of the message. This could be done by getting the user's skill level at log-on time.

Including message Ids It is common to give each message a message identifier or ID. The ID helps software development and maintenance people. But should you display it for users? Yes, in some cases, such as the following:

- In batch messages, adding IDs can help users quickly find error descriptions in a printed manual.

- In online messages that provide only "what went wrong" information, adding IDs can speed up the search for more information in a printed manual.

Never use the ID alone in place of a message. Such "messages" are the terrifying message codes you might have heard about in some systems. Users are expected to understand the code and respond—there is no other accompanying textual information.

If you provide the ID along with the message, provide also the facility for users to turn it off.

The first two or three characters in the ID can identify the product. The last three can be a number identifying the message. You might also want to add one more character to indicate the type or severity of the feedback context. For example, in ABW123, *AB* identifies the project, *W* indicates that it is a warning message, and *123* identifies the message.

10.6.2 CDG 2: conceived from the users' viewpoint

Most of the scarce literature on messages point to the impoliteness or otherwise inappropriate tone of messages. Tone is a problem. But there are lots of polite, nicely worded messages that still fail because they were not conceived from the users' point of view. Similarly, though accuracy of information is critical, an accurate message can be as useless as no message at all if it not conceived and written from the users' point of view.

Take a look at the message in Figure 10.8.

Hole exists.

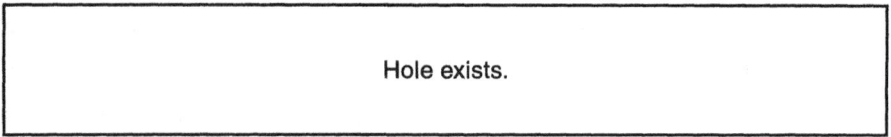

Figure 10.8   A message conceived and written from system viewpoint.

This error message used to appear in a product under development (no longer, thankfully!). Let me hide the details of the software and say that this message appeared when there was an interval between two dates specified by users. In the programmer's vocabulary, this interval is a hole, and thus created the classic message. (The programmer went further to write the following context-specific online Help that was displayed when a frustrated user presses the HELP key: "Hole must not exist.")

Avoid messages such as "Query caused no records to be retrieved" unless you are sure the users who see them are programmers. If you are writing for nonprogrammers, you are better off avoiding terms such as the following:

- Record;

- Row, column, table;

- Fetch;

- Mapping;

- Processing;

- Writing on the database.

10.6.3   CDG 3: context-specific

Give information specific to the message context so that users can respond to that context, and soon continue with their overall task. If it is an error message, communicate what caused the error. If the system cannot identify the actual cause, only then consider providing a list of all possible causes.

10.6.4 CDG 4: clear

If you have written complete, context-specific messages from the user's point of view, you have fully—well, almost fully—achieved design goal four. Here are just a few more points to note.

Use consistent terminology. Also, use the same "phrasing" in similar situations. For example, do not display "Date after which the account should be effective" in one task context and "Date from which account is valid" in another.

In message design, tense is more than just an issue of grammar. Wrong tense can confuse and mislead users. Should it be "was modified" or "is modified"? Get back to the user's task context to find out which is appropriate.

10.6.5 CDG 5: pleasant

Strive to use in messages, qualities seen in good human-to-human communications: tolerance, politeness, helpfulness, finesse. Refrain from hostility. Do not blame, ridicule, or in any way humiliate users.

Excessive praise is also as bad as impoliteness. Avoid praise such as "Wonderful." My little daughter was all excited when she started off with an educational software that repeats "Way to go," "Super de duper," and "Excellent" (in random order) every time she typed the correct response. I could, however, see her getting bored after just an hour of such praise.

10.6.6 CDG 6: attention-getting

With all the graphic and sound techniques available, it is easy to get users' attention. The challenge is to draw attention to a message without being rude, garish, noisy, or ridiculous.

Messages are usually displayed in a standard area on the screen called the message area. The user quickly learns that whatever appears here is a message and that it should be noted. However, consider the seriousness of the feedback when deciding how the message will be displayed. If possible, display the message in a pop-up window (as shown in Figure 10.9) and require a user action (selecting the "OK" button for example) if it can cause damage resulting from a lack of attention.

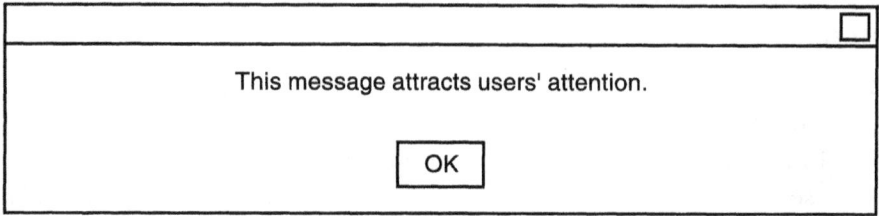

Figure 10.9 Message in a pop-up window.

Sound accompaniment You can use sound, but very carefully, to get a user's attention. Even simple beeps can soon become irritating. Depending on the seriousness of the message, consider providing users the facility to turn off sounds.

Graphic accompaniment Like sound, graphics and color can be effective if used with judgment. Figure 10.10 shows some common graphic techniques.

Other attention-getting techniques Blinking, reverse video, color, "bolding," "boxing," white space, varying typography (typestyle, type size, capitals, italics) are more techniques used to get a user's attention. Use them with caution. Though they are good attention-getters, they can also make text more difficult to read. Other techniques such as underlining and decorating with asterisks belong to the age of the mechanical typewriter. With today's tools, you may not want to go back to those techniques.

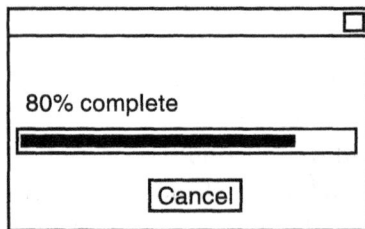

Hourglass   Progress indicator

Figure 10.10 Graphic or visual feedback.

10.6.7 CDG 7: readable

The principles of readability for message display are the same as those for any online information display. For guidelines on ensuring readability, see Chapter 13.

10.6.8 CDG 8: removable

To remove certain informative messages on the mainframe, you must do illogical things such as requesting the display of screen ID (or removing it if already present). Otherwise, if you press the HELP key, you get context-sensitive Help for that informative message, not for the screen!

Removing a message from the screen should be simple. For different types of messages, use different techniques.

Remove informative and warning messages when users take an action that the application can detect, such as pressing the ENTER key, and when the message is no longer needed. For GUIs, CUA recommends certain techniques. Remove an action (error) message only when the application detects that the user has corrected the error that caused the message to be displayed. Remove messages from message pop-ups when the user has pressed any button in the pop-up windows.

10.7 Integration considerations

In each message context, users need various types of information to answer their many questions, such as "What went wrong?" and "How should I correct it?" However, considering the typical restrictions on length, we know that we can only answer one or two questions in the system-initiated message. Fortunately, we have other media such as online Help and printed manuals where we can provide the rest of the required information. The goal is clear: provide all the information that users need. The key is to provide them all at the best possible places, considering project-specific requirements and constraints.

Look at other sections in this chapter for a description of various types of messages. These sections should give you an idea of what types of information may be needed to answer users' questions in each message context.

With project team members, brainstorm questions such as:

- What user questions can we address in the system-initiated message?

- If users need more information, where will that be?

- Are we providing online Help that is context-sensitive to messages displayed?

- Is our client ready to pay for the extra effort and time?

- Does our development environment support easy development of context-sensitive Help?

- Should we only give trouble-shooting information in the printed manual, or will it help to repeat all the messages and their descriptions in a printed manual?

- In which of the printed manuals should we give batch messages and their descriptions?

Answers to such questions should help you decide where, and how, to give all the required information.

References

[1] Galitz, Wilbert O., *Handbook of Screen Format Design*, Second Edition, Wellesley, MA: QED Information Sciences, Inc., 1985.

[2] Duffy, Thomas M., et al, *Online Help: Design and Evaluation*, Norwood, NJ: Ablex Publishing Company,1992.

[3] Nielsen, Jacob, *Usability Engineering*, Cambridge, MA: AP Professional,1993.

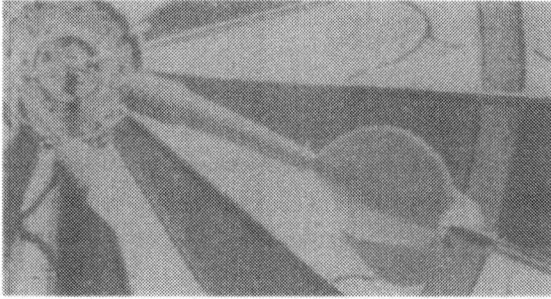

11

Designing Online Support Elements

IN THIS CHAPTER, we will look at some common support elements you may want to design for the online medium. Technical writers can read this chapter to learn how to design Help, tutorials, and other online elements.

11.1 Designing Help

In spite of a "well-designed" user interface, users often get stuck while performing tasks. At times like these, they need quick information, online. One of the things this section helps you in is deciding what Help elements you should provide. It also guides you in designing common online Help elements such as field Help and message Help.

11.1.1 Help and software usability

As they advance from novices to experts, software users could find themselves in various problem situations. One such situation is when a user is stuck with a problem—while in the middle of a task. As Duffey and others [1] call it, the user has "hit an impasse." The task comes to a standstill. The user probably needs to respond to customers, who are waiting in line. Because of the problem, the user's background, and the current work situation (customers waiting), the user may be anxious or frustrated. The user needs immediate assistance—quick support information to aid in completing the task. And the information is required online because the user does not have the time or the frame of mind to go looking for printed manuals.

Software houses have been meeting this user need to a certain extent by providing various online Help elements. Help is quick—and often contextual—support information displayed on the screen, commonly at the user's request. The words "quick" and "contextual" are key features that distinguish Help from other support information elements.

In the real world, you still see many Help systems that are far from being helpful, and those that users never retrieve. Why? Users are probably having trouble. In [2], Sellen and Nicol list the following problems users have with online Help:

- Difficulty in finding the required information;

- Nonavailability of the required information;

- Difficulty in switching between Help and the task;

- Complexity of the Help user interface;

- Poor quality and layout of the information.

No wonder people perceive the software as being unusable.

Although UCID does not explicitly require you to provide online Help, considering the importance of offering quick information online, you will most likely want to do this.

11.1.2   Types of Help

Based on how information is retrieved, we can classify Help as automatic, requested, or retrieved.

Automatic and *requested* Help can be designed as contextual Help. Contextuality is unique to the online medium. Contextual Help is information about user objects (menu options, fields, etc.) or the task the user is *currently* interacting with. This type of Help provides a specific answer to the user's specific question. To get an answer, the user may only have to press a single key, or do nothing at all. There is no need for using relatively time-consuming retrieval devices like an index. In certain types of user interfaces, contextual Help has limited application. In those cases, you can only provide descriptive information rather than procedural information, as you cannot be sure which task the user will perform next.

Retrieved Help is information that is manually found, selected, and displayed by the user through retrieval devices such as index and search utilities.

11.1.2.1   Automatic Help

In automatic Help, contextual information automatically appears when the cursor is on or moved over a user object such as an entry field. No user action by way of explicit key press or mouse-click is required. Contrast this with *retrieved* Help, where the system displays a screen and users have to manually retrieve the relevant topic by making a selection on that screen.

Automatic Help can be displayed in many ways. You can display it in a small pop-up window when the cursor is moved over a user object such as an icon. In commercial products, this is variously called balloon Help and tooltip. You can also display automatic Help "quietly" at the bottom of the screen when the user "tabs" to, say, an input field. In some commercial products, this type of Help is called Hint. Automatic Help is so closely associated with the user interface that you might wonder whether it is interaction information or support information.

Then, there is automatic Help that is intelligent. The system displays contextual information when it "thinks" the user is confused or lost. In some systems, when users mistype a command, information about the format of the command automatically appears. More sophisticated systems "watch" for user performance like repeating an error or a very high error rate. In such situations, the system displays information to correct the user's action.

Automatic help has some disadvantages. As the system must determine the user's intentions (which is very difficult), incorrect "thoughts" about the user's actions can lead to the display of irrelevant information and further user frustration. Some users may consider automatic Help as being intrusive. Others may feel that the computer is in control. This does not conform to the software usability principle, "Users should be in control."

11.1.2.2   Requested Help

In requested Help, contextual information appears when users press a shortcut key, such as F1 provided for that purpose. Contrast this with *retrieved* Help where the system displays a screen, and users have to manually retrieve the relevant topic by making a selection on that screen.

Requested help is the most common type of Help. The advantages are that retrieval is only a keystroke or mouse-click away and the information that appears is relevant to the user's context. You can provide requested Help for each user object and for each type of message. So, you will have Help elements such as screen Help, option Help, field Help, and message Help.

11.1.2.3   Retrieved Help

In this type, Help topics are available. Users must manually find, select, and display topics with Help devices such as an index. This type of Help is similar to a printed manual in terms of lack of context. But it has the online advantage of quicker retrievability. The content can be similar to contextual Help, or it can also be a summary of commands or options available.

11.1.3   Common Help elements

Help may be provided for various user objects, in which case you may have screen Help, field Help, and so forth. It may also be provided to clarify messages, in which case you will have message Help. Each such Help facility you provide is a separate information element because each is likely to be unique in its content and display technique used. The content of each element depends on whether the element is automatic, requested, or retrieved. For example, field Help that appears as automatic Help must be short because of space limitations, whereas the content that appears as requested Help in a pop-up window can be more elaborate.

What follows is a brief description of common Help elements found in a variety of user interface standards and styles. Some, such as screen Help, are more common in a "form fill-in" style of user interface, whereas some, like command Help, are common in systems that support command entry.

11.1.3.1   Screen Help

Screen Help is often contextual information about the screen from where users requested Help. You may provide conceptual and procedural information about an entire screen. You may also want to include information about fields if you are not providing a separate field Help element. See Figure 11.1.

Form fill-in types of user interfaces often provide screen Help, even if they do not provide any other Help element. According to IBM's CUA, screen Help appears if the user presses the Help key when the cursor is in an area for which no contextual Help is available.

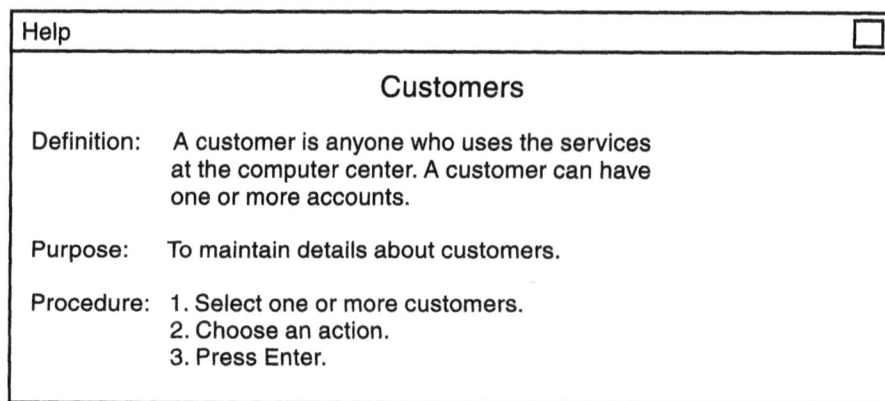

```
┌─────────────────────────────────────────────────────┐ ┌─┐
│ Help                                                 │ │ │
│                                                      │ └─┘
├─────────────────────────────────────────────────────────┤
│                      Customers                           │
│                                                          │
│   Definition:   A customer is anyone who uses the services │
│                 at the computer center. A customer can have│
│                 one or more accounts.                      │
│                                                          │
│   Purpose:      To maintain details about customers.       │
│                                                          │
│   Procedure:    1. Select one or more customers.           │
│                 2. Choose an action.                       │
│                 3. Press Enter.                            │
│                                                          │
└──────────────────────────────────────────────────────────┘
```

Figure 11.1   Screen Help.

11.1.3.2 Field Help

Field Help is often contextual information about the field from where users requested Help. The content depends on whether the field is a selection field, entry field, or output (protected) field. If it is an entry field Help, for example, you may want to mention valid values and defaults. See Figure 11.2.

Field Help is usually displayed when users press a key, such as F1, assigned for that purpose. You can also design it to automatically appear in concise form as a hint.

Form fill-in types of user interfaces often provide field Help. In CUA, field Help appears if the user presses the Help key when the cursor is on a field.

11.1.3.3 Option Help

Option Help is often contextual, "what is" information about the option from where users requested Help. Depending on the design, option Help can also automatically appear. See Figure 11.3.

In CUA, Help for a choice on the action bar (called menu bar in some commercial products) describes the purpose of the choice and how to interact with it. Menu-driven user interfaces and *graphical user interfaces*

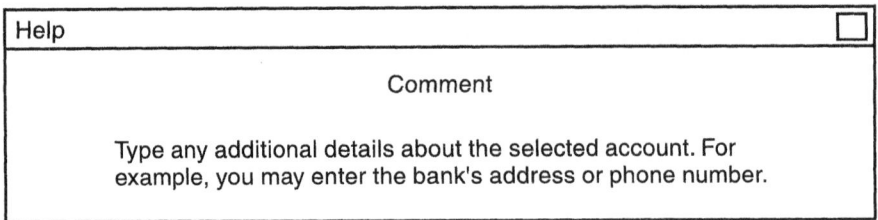

```
┌─────────────────────────────────────────────────────────────┐
│ Help                                                       ☐ │
├─────────────────────────────────────────────────────────────┤
│                          Comment                             │
│                                                              │
│        Type any additional details about the selected        │
│        account. For example, you may enter the bank's        │
│                address or phone number.                      │
│                                                              │
└─────────────────────────────────────────────────────────────┘
```

Figure 11.2   Field Help.

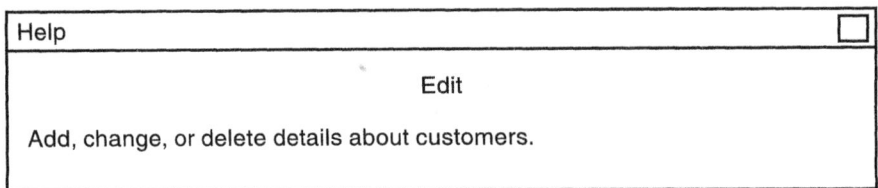

```
┌─────────────────────────────────────────────────────────────┐
│ Help                                                       ☐ │
├─────────────────────────────────────────────────────────────┤
│                            Edit                              │
│                                                              │
│       Add, change, or delete details about customers.        │
│                                                              │
└─────────────────────────────────────────────────────────────┘
```

Figure 11.3   Option Help.

(GUIs) often provide option Help. If users request Help when the cursor is on an unavailable choice, some systems display Help that expalains why the choice is unavailable and what will make it available.

11.1.3.4   Keys Help
Keys Help lists all the application keys and gives the function of each.

11.1.3.5   Command Help
Command Help provides information about one or all commands. In CUA, two kinds of contextual Help are available for commands. If users request Help after typing an invalid command or after typing nothing at all, information about all the application commands appears. If users type a command and request Help, information about that command appears.

11.1.3.6   Hints
A hint, as it is called in some commercial products, is often contextual information about a field or other user object on which the cursor is currently located. Typically, hint is automatically displayed. Content could be descriptive, such as in "Reverses your last action" for the Undo icon. Or, it could be procedural, as this hint in Norton AntiVirus: "Select the drives you wish to scan, then click Scan Now." Write concisely so that the information can easily fit on one line. A hint often starts with a verb, as shown in the above examples.

If an option is currently unavailable, provide a hint explaining why it is not available.

The problem with hints is, users may not notice them as they "quietly" appear at the bottom of the screen. Attention-getting techniques are not used because many users find that annoying.

11.1.3.7   Balloon Help
Balloon Help, or tooltip as it is called in some commercial products, can be thought of as an extension of labels. It is a word or phrase of "what is" contextual information about a user object (see Figure 11.4). In fact, if the cursor is on an icon label that has no text, all that users may get is a text label.

Balloon Help automatically appears when the cursor is over a user object such as an icon. No explicit key press or mouse-click is required.

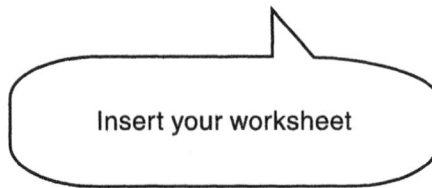

Figure 11.4 Balloon Help (or tooltip).

The information appears in a small pop-up window. The information remains displayed until the user selects the user object or moves off of the control.

11.1.3.8 Message Help

Because of space and other reasons, all the information users need may not appear in the message itself. The information left uncovered in the message must still be available elsewhere. Some companies provide this information as an appendix in a printed manual or in a separate manual if there is a large number of messages. However, as users are in the middle of a task when a message appears, it is a good idea to provide this Help information online—in the form of message Help. Here is an example. For the message, "You could not paste because the copy and paste areas are different shapes," you could provide the message Help shown in Figure 11.5.

If you provide message Help, ensure that a message and its corresponding message Help are integrated.

11.1.4 Critical design goals

By considering the information use model describe in Chapter 6, common Help design problems, and the objective of improving software usability, we can list seven critical design goals for Help, as listed below.

1. Available;

2. Concise;

3. Clear;

Message	☐

The Paste command failed because the copy and paste areas are different shapes.

You did one of the following:

- You selected an area of a table, and then you tried to paste it into an area with a different number of rows or columns.

- You selected text outside a table, and then you tried to paste the text into cells that are not in the same row or column.

Help	☐

Do one of the following:

- Make sure that the area you paste into has the same number of rows and columns as the area you're cutting or copying.

- Make sure that when you paste text into a table, you select only one table cell.

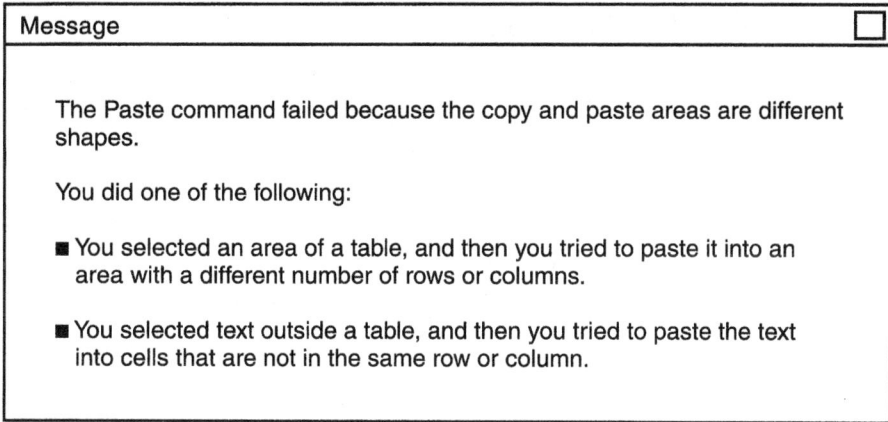

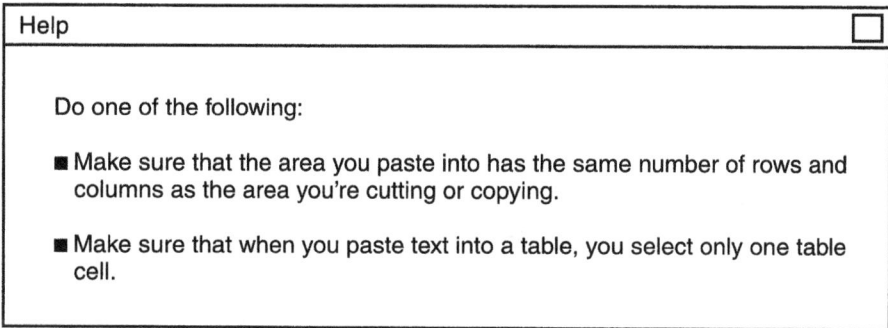

Figure 11.5 Message Help.

4. Retrievable;

5. Contextual;

6. Usable;

7. Readable.

Whereas the first three goals are related to content, the last four are about interaction. The goals set assumes accuracy, and therefore does not list that item. If you create entirely new types of Help elements, you may want to tailor a new set of goals.

11.1.5   Designing the content

Now, the aim is to meet the three critical design goals related to content: available, concise, and clear.

11.1.5.1   CDG 1: available

The reason why Help elements appear in your information architecture is because you want to quickly answer specific user questions. To ensure completeness, you should first know the questions users have in their minds when they seek Help. Their questions often determine the type of information they need. Table 11.1 lists typical questions in users' minds—while they are performing tasks:

Use Table 11.1 to determine the types of information users may need. Read the section, "Integration considerations," to learn how to integrate Help elements with other information elements.

11.1.5.2   CDG 2: concise

Conciseness is critical for information that is targeted for display on the screen. Any information that is not required *now* will waste the user's time. So, irrelevant information is out. And that calls for "chunking" of information into meaningful topics, and perhaps layering of information, so that users see only the topics they need at the time.

Table 11.1
Users' Questions and Information Types

Question in Users' Mind	Type of Information
What is this?	Definition
How do I use this?	Procedure
What happens if?	Effects
Is there a better way?	Advice
What went wrong? Why did this happen? Why can't I?	Troubleshooting
Is it mandatory? Is there a default value?	Other

"Chunk" topics Do not think of a page or a screen as the basic unit of information. What do users think of as a unit and want to see at one time? Topics. A topic can be a definition, a procedure, or an answer to one specific question. If the information you have does not relate to the question being answered, put it elsewhere. Here is an example of "irrelevant" information. While writing procedures, if there are many methods, you only need to provide the simplest and most common method. If users want know about other methods, you can provide links to that information. An independently accessible topic, however, should be complete in itself to avoid "jumping" the user to other topics.

Layer information Consider layering your information. Layers can serve different users or different purposes. A related technique is called elaboration: providing layers of information that users can choose to access—either for more or less detail. Layer information moderately.

Write tight Write tight, short sentences. Watch out for long sentences and paragraphs. They are hard to read, understand, and use. Try tables and lists instead.

11.1.5.3   CDG 3: clear

Clarity is achieved through many factors described in Chapter 13. Concise writing, a critical design goal for Help, is one of them. Consistency, another factor that contributes to clarity, should be highlighted here because of its importance in online communication.

Use terminology consistent with the product. Use parallel structure—the kind of template approach you see in reference manuals. Have similar text for similar things. For example, if you say "Date after which the account should be effective" in one place, do not say "Date from which account is valid" in another. Use tables and lists wherever possible.

11.1.6   Designing the interaction

Make getting help so simple and dependable that it becomes second nature [3].

The Help system is like an application by itself. However, the Help user interface must be consistent with the rest of the product user interface. It should be so easy to use that it becomes a natural part of using the software. If users must learn to use Help, you are defeating the purpose of providing it in the first place.

11.1.6.1   CDG 4: retrievable

Consider providing the following retrieval devices:

Visible   Users should be aware that text-based Help is available, so have it always visible, perhaps through an icon or option. Make sure it consistently appears at the same place. Users should also know what types of Help are available.

Shortcuts   There are two types of shortcuts. First is the shortcut for requesting Help. Define a key (or button) for this purpose. Also, respond to "natural" choices such as ?, Help, and F1. If you are layering information, instead of requiring users to pass through the layers, provide shortcuts to specific facts.

The second type is the "Do it" shortcut that automatically performs a step. For example, it can automatically open an object so that the user does not have to remove Help and then start looking for that object. Such shortcuts make users more efficient. They also reduce the amount of information you need to provide.

Contents   This is a list of topics organized by category. In Windows, a Book icon represents a category or group of related topics, and a Page icon represents an individual topic. To display a topic, users select a topic and click on the Display button.

Index   The index is a list of alphabetically sorted labels, tasks, concepts, abbreviations, symbols, and so forth related to the software system. Users can search the index to quickly find the information they need. Consider providing alphabet buttons with hypertext links to other parts of the Help window. This avoids scrolling to first find the alphabet letter then the required word or phrase. In Windows Help, users enter a keyword or select one from the index then click on the Display button to display the selected topic.

Search   This is a facility whereby users specify a topic they want to retrieve from the Help file. If you are providing a keyword search facility, try and include a good set of search keywords. Most Windows applications have a very short keyword list. For most users a full-text search is more useful than a keyword search facility.

Hypertext navigation   A hypertext jump (or hyperlink) is a button or interactive area that takes users to another topic. Provide a visual cue for the presence of a hypertext jump. You can do this by underlining the words or by using a different font or color. In Windows, the default system presentation for text jumps is green underlined text.

Do not try to link everything to everything else. Keep the link path simple and logical. If you have too many hyperlinks on a page, consider keeping them all in one place—perhaps under "See also" at the end of a topic.

Other navigation devices   Other devices you should consider providing are backtrack, cross-reference ("See also"), bookmarking, history, previous and next buttons, Go To, index, search, and scroll bar. Some of these devices are described in Chapter 13.

11.1.6.2   CDG 5: contextual
Display information relevant to the user's current situation or question. Most types of user interfaces do not support access to procedural information that is contextual. They only display descriptive (conceptual) Help for user objects. This is because in a context, several tasks may be possible. In such cases, what you can do is display a list of relevant topics from which the user selects.

11.1.6.3   CDG 6: usable
There are two common design techniques that trouble Help users:

- Help appears in another screen, completely hiding the application screen on which the user was working. This technique requires the user to switch back and forth between the application screen and Help screen.

■ Help completely disappears when users try to apply the information on the application screen. This requires users to memorize or note down the information before they leave the Help screen. Worse, it requires them to retrieve Help again if they forget what they read.

With these design problems, most users will stop using Help. Help must be *used* easily and quickly if it is to improve software usability.

Users should be able to read the information and use the application at the same time. The information should be available as long they want it.

Both application task and information (especially procedural information) must always be visible. Let Help stay on top, even on top of modal dialog boxes. Consider a "floating" window, which is modeless and is always visible while the application is active. Consider displaying Help in variable locations—that is, above, below, and to the side of the object for which Help is requested. Depending on the type of user interface, you can also consider a fixed location on the screen where you will always display Help.

Size and place the Help window so that it covers minimum space. But, make sure it is large enough to allow the user to read the information, preferably without scrolling. Avoid excessive hyperlinking and having too many Help windows open at the same time. For removing Help, consider responding to users' "natural" choices such as ESC, Quit, Exit, Q, and X.

Make Help consistent with the rest of the software. Do not turn it into its own complex application.

11.1.6.4 CDG 7: readable

Make the Help window large enough to let users to easily read the information. Use simple formatting. Avoid italics except for headings. Indent text in tenths of inches rather than in quarters or halves. Handle line spacing with care. Increased line spacing can reduce readability. Limit the use of color in text—it can be distracting. Using colored text usually means "hardcoding" the background and text colors. Rather, you should allow the user to control such subjective characteristics. Using colored icons, on the other hand, to mark information helps users identify the kind of information presented.

Make creative use of icons to improve visibility and readability. The Windows 95 Help browser has interesting icons for topics and categories. For example, when users click on a Book icon, the icon changes into an open book.

11.1.6.5 Other interaction design goals

Make Help customizable by users. The facility to customize is useful if Help is for a large product that has a wide range of users. Users at each installation may want to add, modify, or delete Help information. Besides the facility to customize content, users should also be able to specify how they want to interact with Help. This is at the individual user level. Such customization can include things like changing the size, shape, and colors of the Help window.

11.1.7 Integration considerations

In some software products, the message box also includes a Help button. Here, message and Help kind of coexist as you can see in Figure 11.6. How important it is to integrate! Imagine two independent teams working on messages and Help, not knowing what the other is writing.

Help elements are key contributors to software usability. What Help elements will you provide? Providing all the ones listed earlier in this chapter might not be beneficial. You need the right mix. Therefore, integrate! Integrate all Help elements and integrate them with the other information elements. Will screen Help repeat field entry details if you

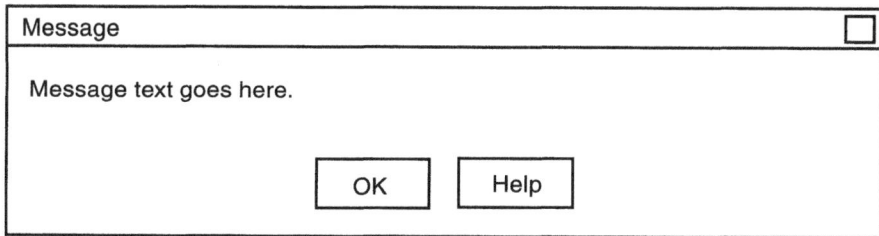

Figure 11.6 How important it is to integrate when they are this close!

also provide separate field Help? How about Help for output fields? Can procedural information be contextual?

Consider the software's user interface thoroughly while deciding what Help elements to provide. Here is a small example. If the user interface shows icons without labels, you may decide to provide balloon Help that gives the textual label. If the user interface shows icons with labels, you may still want to provide balloon Help, but give a little more elaboration on the label. When you integrate like this—based on information requirements—you are actually integrating all the information elements with the rest of the software. And that is great news from the users' point of view.

11.2   Designing tutorials

The online tutorial is an instructional course designed with user interaction. The goal of the tutorial is to help users feel confident about two things: the software system's capability to help do the work and the user's own ability to use the software system. Toward achieving this goal, the tutorial should familiarize users with the software system's features, its user interface, and the procedures for basic tasks.

The primary users of a tutorial are novices. Secondary audiences can be occasional and transfer users. For all these groups, the tutorial provides a safe environment to try out the system.

Depending on your project-specific requirements, there are many types of tutorials. *Paper-based tutorials* are the easiest to develop. *Online tutorials* have the advantage of interaction. Users can try the software system in a safe, controlled environment. However, online tutorials also have the disadvantages associated with information presented online. A poorly designed online tutorial is worse than no tutorial at all. Online tutorials that are just electronic page turners, where users read information just as they would from a book, and those that have a complicated user interface, are not useful. Finally, you can also consider an *online-paper combination tutorial*.

Sometimes, tutoring is provided via video and classroom training. Whereas video-based tutorials lack interaction, classroom training is often a one-time affair and repeating it can be expensive.

The most effective tutorials are those that are online and interactive. However, from a development viewpoint, they are expensive. In this section, we are going to look at designing online interactive tutorials.

Developing an interactive tutorial is a small software development project by itself. You need to plan the effort thoroughly, making sure you have all the required resources. Here are the activities related to designing an online interactive tutorial:

1. Identify and organize lessons;

2. Decide standard content of lessons;

3. Decide on interaction techniques;

4. Design tutorial's user interface;

5. Write lessons.

11.2.1 Identify and organize lessons

Typically, online tutorials consist of one or more modules, each having two to four lessons. You should decide on each lesson based on target users and their learning goals.

Lessons can be simple tasks, more complex tasks (for example, to get users to use other information elements), a demonstration of features, or important concepts. Often, the tutorial is narrowly focused—explaining only the basics.

The *simple-to-complex* organizing approach is frequently used for tutoring. Another approach that has proven to be effective in learning is the *known-to-unknown* approach. Here, the belief is that people learn about the new in terms of the familiar.

In general, arrange lessons for users to build on knowledge learned in earlier lessons. Do not organize alphabetically. People do not learn in alphabetical order.

11.2.2 Decide standard content of lessons

Consider ways to achieve your instructional goals. What are the standard things you should have in the tutorial? The following are frequently used.

- *Lesson objective.* Identify the topic and the target users. Make it clear what each lesson is about and what users will learn. List the expected result, which could be acquiring conceptual knowledge or procedural skills. Wording should motivate! Explain how the information or tutorial will help them.

- *Time.* Mention the approximate time it will take to complete the lesson. Lessons should be designed so that users won't spend more than 30 minutes to complete each one.

- *Task concepts.* Briefly describe task concepts by answering typical user questions such as why the task should be performed. Refer to the task information checklist given in Chapter 12. Consider providing a scenario. Scenarios are task situations that typically appear in the real world where the product is used. An example senario: "You're going to prepare a budget to see if you can afford to buy a home."

- *Task procedure.* Give the steps required to complete the task. If steps go over eight, try and reorganize into subtasks.

Start with a lead-in phrase like "To perform the xyz task, do the following." Number the steps sequentially. Detailed steps may be required if users are new to the operating system or computers. For example, you may have to say not only *what* should be selected, but also *how* in terms of mouse or cursor movements.

Steps should be generic, listing the detailed procedure users would follow every time they perform this task. For example, "Specify the response time in the Time field" not "Enter 7 in the Time field." If 7 is shown entered in the screen shot, add another sentence like "Here, we have entered 7." Combine steps if appropriate and if it is not confusing. For example, "Select File Open" is preferred to "1. Select File 2. Select Open."

If required, explain or graphically show the system response. Consider showing screen shots. You may want to add callouts to screen shots indicating which button to click or option to select if it is not easy to locate, especially for options labeled with icons.

- *Examples*. Consider providing examples that reinforce or clarify what is said in the text. You can also provide other kinds of examples that could be separately accessed. In the latter case, do not repeat the conceptual information already given. Finally, make sure that all examples are real-world, working examples.

- *Summary*. It is a good idea to repeat the most important points from the lesson. Avoid the traditional "You have learned how to...." Also, avoid repeating objectives or lengthy descriptions.

- *Next*. Suggest what to do or where to move next.

Besides these methods, you might also want to have one or more of the interaction techniques (see Section 11.2.3) as standard items for each lesson.

11.2.3   Decide on interaction techniques

Based on your instructional goals, decide on the interaction techniques you should provide. Some interaction techniques to consider are demonstration, handholding, and practice session.

- *Demonstration*. The "See how I do it" feature. Try to provide a self-running demonstration in your tutorial to show product features and functions.

- *Handholding session*. The "Come, let's do it togther" feature. Instruct at every step and let the user perform the actions. Do not leave anything for the user to figure out. Save time and clerical work for the user by filling up fields for them. Anticipate every move and provide feedback.

- *Practice session*. The "Go ahead and try it yourself" feature. Hands-on for users to try out what they have learned. Practice is very important because users should get familiar and comfortable with the user interface. Provide short, frequent practice sessions.

Ask a user to perform a particular task. This is like actually using a portion of the software system itself. The advantage is that users can make errors without suffering the consequences. Design the practice session in such a way that there can be no damage. Warn users wherever you cannot entirely eliminate damage.

In a practice session, you want users to perform an example task, so ask users to enter example values. Therefore, unlike your example in task procedure, you say "Enter 28" rather than "Enter a value." Also, do not handhold as in "Enter 28 in the Temperature field." Instead, say "Specify a temperature of 28 degrees Celsius." Let users learn where to enter the information. Keep some Help available for those who cannot find the field!

Give feedback that is specific to the user's action or the system's response. Explain whether the user action was correct or incorrect. For correct action, you could say something like, "You have successfully added the required information. Click the Back button to return." For an incorrect action, you could say, "Your entry was not correct. Please select Retry to try again or Cancel to continue." Take care not to offend or humiliate the users.

11.2.4   Design tutorial's user interface

An online tutorial I recently evaluated was as difficult to use as the software system it was trying to teach! A tutorial's user interface must be relatively friendlier. Keep it simple and design it for ease of learning. Otherwise, you defeat the very purpose of providing a tutorial. Consider the following:

- *How will it be invoked?* Make it easy to start up. Is it separately packaged in a CD-ROM or diskette? Will it be available from the Help menu?

- *How will it be used?* Make it easy to use. Instructions on how to use the tutorial may still be required by users who are lost. Make such instructions available on request.

- *Navigation.* Decide how users will navigate the tutorial. Provide a lesson menu. Number the menu options if lessons should be taken in sequence. Mark the lessons already taken. While using a tutorial, users should always be able to display the lesson menu.

Users should be able to begin a new lesson, end the tutorial, or choose topics in any order. Do not change displays automatically. Some people read more slowly than others. Faster readers feel trapped.

Provide alternative paths through the information based on user choice. Allow backward branching through a simple action. Allow sideways branching to display extra information. Also, allow users to jump ahead—that is, move forward in the tutorial. Some users feel trapped without this jump ahead facility.

- *Feedback.* Feedback is most important in a tutorial. Respond within the expected response time—neither too late nor too quick. Respond consistently. See Chapter 10 for more details on designing feedback.

- *How will it be closed?* Make it easy to close. Consider providing exits on every screen.

11.2.5 Write lessons

Each screen should present a single but complete idea or topic. Talk to the user as if you were sitting right beside him or her, offering help. Write short sentences and paragraphs. Use lists wherever possible. Users are not interested in screen after screen of information. Find every possible way to build interactivity and hands-on experience.

Use interaction wherever possible. Provide additional explanatory or hint text via *pop-up* windows and hypertext. Use a pop-up to describe the area of screen on which the user clicks. You should indicate the availability of a pop-up in the text. Describe what is available and how to display it. For example, you could say "Click on each option for more information." Use hypertext to define a word or phrase. Visually indicate, maybe by blue underline, that the word or phrase is hypertext. Provide definition of the term clicked. Do not overuse.

Design screens and user objects so that they guide the user's eye movement. Design a layout that allows enough space for graphic images and for users to see all the information without scrolling (you can use hypertext pop-ups to provide more information). Basic layout should be consistent, but you can switch between a *side-by-side* and a *top-and-bottom* approach for text and graphics. See Figure 11.7.

Side-by-side layout

Body text goes here. And here and here. Body text goes here.

And here and here.Body text goes here. And here and here. Body text goes here. And here and here. Body text goes here. And here and here. Body text goes here. And here and here.

And here and here.Body text goes here. And here and here. Body text goes here. And here and here. Body text goes here. And here and here. Body text goes here. And here and here.

Graphic goes here.

Top-and-bottom layout

Body text goes here. And here and here. Body text goes here. And here and here. Body text goes here. And here and here. Body text goes here. And here and here

Body text goes here. And here and here. Body text goes here. And here and

Graphic goes here.

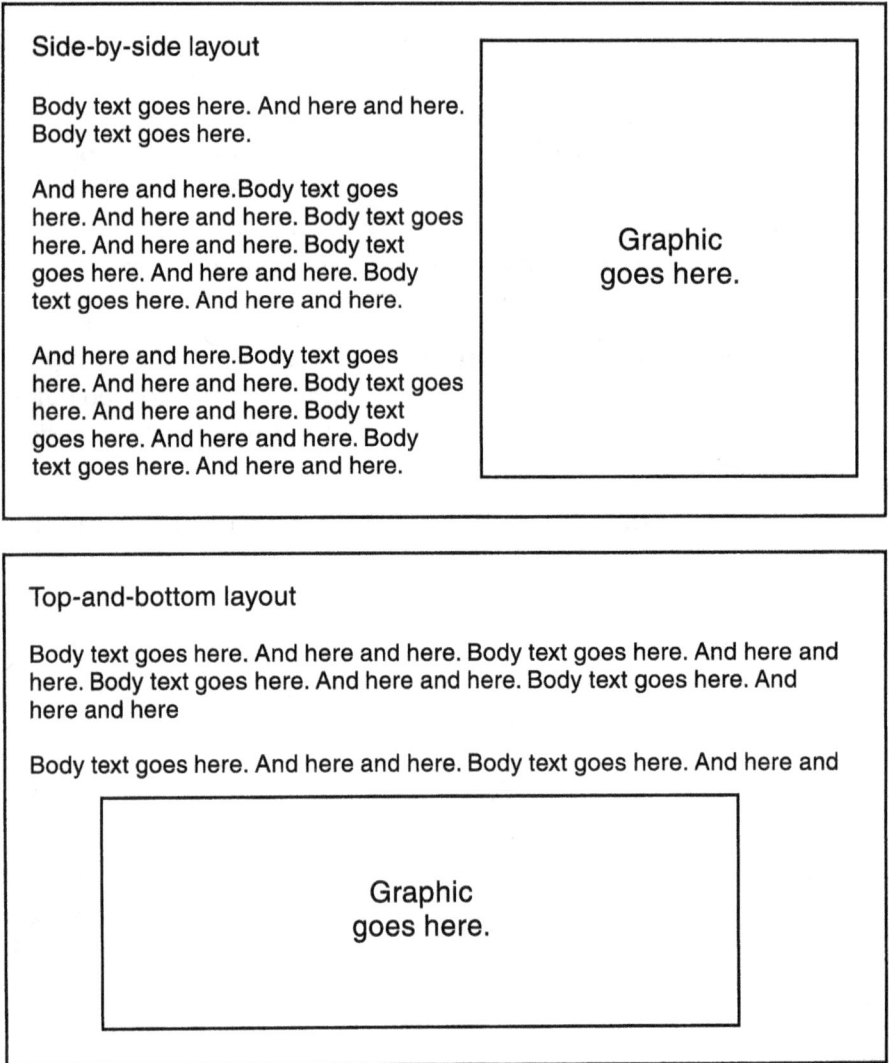

Figure 11.7   Tutorial lesson layout.

Avoid filling an entire screen with text. Try and have at least one graphic image for every two screens. However, every graphic image should have instructional value. Choose from screen shots, screen cuts, illustrations, and photos. Make sure screen shots show all the buttons and other screen objects mentioned in the text. Graphics should be friendly, not technical. Use illustrations and stock photos to convey how the user might relate to a concept. For example, you can show a clock to illustrate the ideas of elapsed time. But avoid obscure visual metaphors. Be sensitive to localization and internationalization issues. Avoid cliches and culturally biased images.

Use callouts to draw user's attention to a specific area on a graphic. Callouts are short text—a word or phrase in small boxes. Here is an example of a callout: "Click here to display Printer settings."

Rather than a still graphic, consider animation to effectively illustrate, say, a process. You can use animation to show, for example, the flow of goods through a supply chain. Provide the facility to turn off animation.

11.3   Designing other online support elements

Today, software houses are increasingly providing "more efficient ways of doing things" for users. The "advice" information you find in some products may even include application domain tips. For example, Harvard Graphics offers Design Tips that "make it easy to give your presentations that fit-and-finish look, even if you aren't a graphic designer," as the product's advertising claims.

It is now becoming common to make Help interactive by providing optional demos and examples wherever appropriate. Moreover, multimedia that began in online tutorials is being tested in the online Help area. And Help retrieval by voice activation and gesture are also being researched.

In this section, we will look at wizards, cue cards, online manuals, and readme files.

11.3.1   Wizards

A wizard is a tool that – through interaction—helps a novice user through a task. It includes controls and default settings to gather input from the user. The input and the navigation devices, such as Back and Next buttons, help users easily complete the task.

Wizards supplement, rather than replace, the facility to directly perform a specific task. They are typically designed to hide many of the steps and much of the complexity of the task being automated. They are especially useful for automating infrequent tasks that users may have difficulty learning or doing. Wizards allow the user to be in control of the task the wizard automates.

Wizards are not really considered support information. They operate on real data. We talk about them here because they (1) support tasks and (2) have instructional text such as "To begin creating your letter, click Next." However, most instructions in wizards appear in the form of questions—for effectiveness. For example, say, "Which of the following styles would you like to use for your presentation?" rather than "Select a style from the following list." The text is concise and self-explanatory. A wizard is designed to make a user's completion of a task easier. Therefore, there is no need for support information such as Help.

TipWizard   The TipWizard you find in some Microsoft products provides brief information about better ways of doing things. (Similar support elements are Borland's Experts and Lotus' Assistants.) Hiding in the background of the software system, the TipWizard is activated when it notices a task that could be performed more efficiently. When this happens, the light bulb on the TipWizard icon bar "glows." Click on the icon and a shortcut such as this comes up: "To insert the current date, press ALT + SHIFT + D." Users who prefer to strike keys rather than wade through menus will appreciate this tip and the facility. TipWizard suggests related and new features. It also shows a tip each time users start the application, but offers the option to turn it off.

11.3.2   Cue cards

Cue cards are aimed at helping users quickly complete tasks online. Each cue card includes step-by-step procedures for a task. In Microsoft Project, the cue card "floats" above the project window to allow users to read and follow the instructions at the same time. To use cue cards, users choose Cue Cards from the Help menu.

WarpGuide, IBM's intelligent task mentor, uses a SmartGuide role for the task of system customization. The amount of Help is automatically determined based on the user's previous actions. For the "Finding a file" task, here is the Help:

Novice: "Type the name of what you want to find."

Intermediate: "Type the name of the file, folder, or object. To find files of one type, use an asterisk; for example, *.BMP locates all bitmap files."

Advanced: "Type as much of the name as you know, and use an asterisk for the rest. Example: Type my* or *file, to find things that begin with *my* or end with *file*, respectively. Use an asterisk for any parts of a name that you don't know."

Expert: "Tip: If you want a new look for your icons, search all drives for files that end in ICO. When the collection of icon files displays, open Properties for the object whose icon you want to change, click the icon, and then drag the new icon file to anywhere on the Icon page."

These cue cards show different kinds of information (such as definitions and examples) and different levels of information (for novice, etc.). The COACH agent selects the kind and level based on a user's history, and the cue card information button gives access to information of all kinds and levels.

11.3.3   Online manual

The online manual is the electronic equivalent of a paper-based manual. With its lack of contextuality, it may not be very helpful if it is the only online element users have access to. In some software systems, it is used as the source material for use by a Help system that displays contextual information. In others, it is meant to be printed and used. The online manual is often designed like a printed manual in terms of content and organization. Of course, it has the online advantage of faster retrievability.

11.3.4   Readme file

This is a file with *readme.txt* as file name. Topics you can provide include a list of files going with product, installation instructions, technical tips, and corrections to information elements. The readme file contains concise, self-contained information. Text should be unformatted for easy viewing on any word processor or editor and for quick printing.

References

[1] Duffey, Thomas M., et al., *Online Help: Design and Evaluation*, Norwood, NJ: Ablex Publishing Corporation, 1992.

[2] Laurel, Brenda, *The Art of Human-Computer Interface Design*, Reading, MA: Addison-Wesley Publishing Company, 1994, pp. 143-153.

[3] Horton, W., *Designing & Writing Online Documentation*, New York, NY: John Wiley & Sons, Inc.

12

Designing Printed Support Elements

IN ADDITION to helpful online information, users often need detailed information on paper. In this chapter, we will look at support elements meant for the print media. Technical writers can read this chapter to learn how to design common printed manuals.

12.1 Printed elements and software usability

Before attempting to install a product, users may want to read the installation instructions. The print medium is often the only medium for such information. In the case of large and networked systems, a lot of planning may precede actual installation. Print medium is again best for such detailed planning information. Prior to performing an end-use or support task, users may want to know the concepts relating to that task. *Conceptual information* answers questions such as why and when a task should be performed. Printed elements are always best for providing such conceptual information. In all the above cases, the user needs information that could be hard to read on the screen—but often has the time to look for and read support information. Printed elements are the way to address this user need.

In the real world, we see printed elements that are far from being helpful and those that users never use. Possible reasons are the lack of conceptual information users seek, bulkiness of the printed element, difficulty in finding the required information, and poor layout of the information element. People therefore perceive the software as being unusable.

12.2   Common printed elements

UCID does not demand that you provide printed information elements. However, you are very likely to provide printed elements because of the need—on paper—for details and for conceptual information. Let's look at some common printed elements.

Getting started   Printed information provides users, especially novices, information for initial product setup and use. This can be quick installation instruction that does not go into advanced installation and customization tasks. Other examples of the type of information that is usually covered in a getting started manual are: procedure for starting and quitting the software system, a quick exploration of the software system's features, and roadmap to various support information elements. In many software systems, getting started only appears as a part or chapter in another information element such as a user's guide.

Tutorial   Tutorial users can be fearful novices or simply I-want-to-learn-before-I-try users. The tutorial helps them safely "rehearse" a few basic tasks before they actually perform on the job. Users may need this element to ease the shock of a new software system's perceived complexity. For small software systems, one or two sessions may appear as part of another printed element such as a user's guide. The trend, however, is to move tutorials online and to make them more effective via interactivity.

Quick start   This gets experienced users going fast. It is especially useful for a new version of a software package. Microsoft Word has a *Quick Results* manual. It describes how to install the product, explains what's new in this version, and shows how to use wizards to create documents.

User's guide   Covers conceptual and procedural information about end-use tasks. If the product requires only a few support tasks, information about them may appear here, too. Otherwise, such information typically appears in a separate guide such as an administration guide. Some organizing techniques you should consider for a guide are: chronological, frequently-used-tasks-first, and simple-to-complex.

Reference manual   A reference manual may be required if you feel that some information is used by most of the users, most of the time. Such information is often a structured description of commands. Although it provides complete explanations, a reference manual often functions as a reminder. Controls, language, troubleshooting, and technical specifications are often packaged as reference manuals. Aimed primarily at experts, the reference manual is used *occasionally* for a short while. Typically, it consists of topics. The topics are almost always organized alphabetically. Everything about a topic is in one place. And this is important because reference information is often read out of context. A command language reference would be structured alphabetically, showing the complete syntax and semantics for each command.

12.3   Writing task information

People buy and use software to perform tasks. Tasks, therefore, are at the center of software usability. We have seen that tasks can be end-use tasks or support tasks. While describing them, you need to provide conceptual and procedural information.

Conceptual information   As mentioned earlier in this chapter, conceptual information answers questions such as why and when a task should be performed. In the user's guide for a project management software package, conceptual information can be a page or two about project management. Here, you might define project management and describe project management activities like planning and resource allocation. The following is a checklist of questions you may need to answer.

- What is the task?

- What subtasks comprise the task?

- Why do this?

- When to do and why? When not to do and why not?

- Who should perform it?

- Can several users perform it at once?

- Who will use the information?

- What are the prerequisites?

- Is it required or optional?

- How frequently?

- What mode of processing is recommended?

- How do I do this?

- What is the effect or output?

Here is sample of conceptual information for the task of computing revenue using an accounting software package:

> Once you have generated a price list, you may want to test the effect of the prices on the revenue. You can do this by applying the prices to one of these: forecasted usage data and historical usage data. The process of computing revenue is known as simulation.

Performing simulation helps you evaluate the revenue and profit against your company's financial performance objectives. If you find wide variances, you can change the prices in the price list and perform simulation again. You can repeat this process until you are satisfied with the simulation results. Though it is not mandatory, simulation helps you to correct and set competitive prices in your company.

Procedural information   Procedural information is more than just step-by-step instructions. For large-scale tasks, you may want to provide an overview. If the number of steps is large, give a big picture view of the steps or break them into subtasks. Here is an example of procedural steps:

To install a printer driver:

1. Select one or more printer drivers from the list.
2. Select Install.
3. Ensure that the information in the Directory field is correct.
4. Select OK.

The level of detail depends on the purpose and the target users. Consider if you should also give the purpose of the user action. For example, "Select Find" is a user action, whereas "Select Find to locate the document" adds the purpose of the user action. In tutorials, you may also want to say how to perform the user action.

A detailed commentary of what goes on between the user and the computer may not be required except in tutorials. However, you may want to show important screens or dialog boxes, describe important fields, and mention unusual system responses.

Number the steps. Start each step with an imperative verb where possible. Use similar phrasing for similar steps. You need not strive for consistency in terms of parallel structure. For example, you can write "From the File menu, choose Add File" when other steps start with an imperative verb such as "Type..." or "Select...."

After the steps, describe what has been done and what users are now ready to do.

12.4 Writing reference information

When it comes to presenting reference information effectively, you need to focus on two factors, organization and structured writing. Both of these factors are aimed at improving the retrievability of information.

Organization of reference topics is usually alphabetical.

Structured writing is essential for reference information (see Chapter 13). Topics are clearly defined and a standardized structure is used. Similar items within topics will have the same headings for improved retrievability. Users may only be looking for specific information, such as the syntax of a command. Structured writing allows a fast lane to exactly that small piece of information the user is seeking. It eliminates the need to read chunks of paragraphs to find the required information.

In *Visual Basic Language Reference* [1], the standardized structure has these headings: Applies To, Description, Usage, Remarks, Data Type, See Also, and Example. Figure 12.1 shows how this structure is used for the reference topic called GridLineWidth Property.

12.5 Integration considerations

Integration is the most fundamental requirement in information design. Identify information elements and integrate them. How will the printed elements tie in with the other information elements? Will they complement? What is the rationale? While deciding what information to cover in which element, consider the user's characteristics, their need for information, and the model of information use. The following are some points for you to consider.

Generally, support information is categorized as tutorial, guidance, or reference information. Tutorials increasingly appear online. Guidance information appears in user's guide, quick start guides, and so on. Reference information such as descriptions of commands typically appears in a reference manual and in reference cards. You can package a printed element to contain one, two, or even all three of these types of information.

Remember that in the case of large and networked systems, certain support tasks such as installation and administration are done by people other than end users. Support tasks that can be performed by the

Applies To	Grid.
Description	Determines the width in pixels of the gridlines for a grid control.
Syntax	[*form.*] *grid*.**GridLineWidth**[= *numericexpression*]
Remarks	The minimum setting is 1; the maximum setting is 10. The default is 1.
Data Type	Integer
See Also	GridLines Property.

Example	The example increments the width of gridlines in a grid with each click of the mouse. For information on running this example, see Help.
	
	Sub Grid1_Click ( )
	
	If Grid1.GridLineWidth 10 Then
	
	   ' Increment width.
	
	   Grid1.GridLineWidth = Grid1.GridLineWidth + 1
	
	  Else
	
	   Grid1.GridLineWidth = 1 ' Set width back to 1.
	
	  End If
	
	End Sub

Figure 12.1 Structured presentation of reference information.

product's end users, especially if the information is less, may appear in a printed element such as the user's guide.

Software systems that have form fill-in types of user interfaces, which are primarily used by data entry clerks, often need a screen-oriented approach to presenting information. You may want to show each screen and explain what users can do in the screen, and how. This approach may require field information if online field Help is not provided. In graphical user interfaces, you may not want to describe every menu option. Because of the availability of windowing and direct manipulation capabilities, you can provide "what is" information about menu options on the screen rather than in a printed element.

Make sure that explanatory information about messages is provided. If this cannot be presented online for some reason, you should plan to provide such information in a printed element. You can provide this information as an appendix or part in a printed element or in a separate messages manual if the number of messages is large.

Reference

[1]   *Microsoft Visual Basic Language Reference*, Microsoft Corporation, 1993.

13

Achieving Information Design Goals

THIS CHAPTER provides techniques that you can use to achieve information design goals for various information elements. The complete list of information design goals (ARRCUPA) is in Table 13.1. Technical writers can use this chapter as a mini-reference. For example, if you are concerned about the retrievability of an information element, check here for various techniques to improve retrievability.

13.1 Availability

All information users need should be obtainable from one of the information elements provided. One approach to deciding on what to provide is to look at information from the viewpoint of task performance. If any

Table 13.1
Information Design Goals: ARRCUPA

Available
Retrievable
Readable
Clear
Usable
Pleasant
Accurate

piece of information helps in performing a task, include it. Use the list below to see which topics are relevant in the context of your project.

- What is this task/user object?
- What subtasks comprise the task?
- Why do this? Why use this?
- When to do and why? When not to do and why not?
- Who should do this?
- Can several users perform at once?
- Who will use the information?
- Is the task/action required or optional? Is there a default value?
- What are the prerequisites?
- What mode of processing is recommended?
- How frequently should the task be performed?
- How is it done?
- What is the effect or output?
- What went wrong?
- Is there a better way?

13.2   Retrievability

We know that users want to complete their tasks quickly and with as little effort as possible. Meeting these user objectives is hard if users also have to read support information. Further, the harder and slower it is to get to the required information, the fewer the users who will get it. The implication is that we need to design support elements so that people can locate information quickly.

Retrievability refers to how quick users can get the information they are looking for in an information element. How can we minimize the lookup time? We can provide table of contents, an index, and other retrieval devices. The more retrievable devices the better. That's because different people may prefer to use different retrieval devices. We should also design each device well. Online retrieval is faster and more efficient. For the online medium, you also have more retrievability devices such as context-sensitivity, search/query, hypertext, and visibility.

13.2.1   Attention-getting devices

Before users can retrieve or read information, they should be aware of its availability.

Always keep users aware of the availability of various online support elements. For example, show a Help button if Help is available. Use attention-getting devices to let users know when you display a message. One way to get attention is by displaying the message in a pop-up window that requires a user action such as the selection of an "OK" button.

With all the graphic, sound, and other techniques available, it is easy to get users' attention. The challenge is to draw attention without being rude, garish, noisy, or ridiculous.

13.2.2   Appropriate organization

Good organization reveals how pieces of information fit together, emphasizing main topics and subordinating secondary topics. It does not force users to jump to other locations. Good organization is important even at lower levels—that is, the sequence of paragraphs and sentences is important, too. Besides improving retrievability, good organization enhances the clarity of information. Careful organizing is perhaps more critical in printed elements, where users can see the overall organization as they scan the contents or even as they casually turn pages.

Some information organization approaches are described here. All these approaches can be used at various levels in both print and online media.

- *Chronological.* Chapters that describe tasks, in a user's guide, for example, are often organized based on the order in which the tasks are expected to be performed.

- *Frequently-used-first.* The tasks more frequently performed are described first. This is why in some manuals you see installation—a one-time task—described toward the end of the manual though it is the first task.

- *Known to unknown.* In this approach, the belief is that people learn about the new in terms of the familiar. As it promotes learning, this approach is often used in tutorials.

- *Simple to complex.* Again, this is an organizing approach that promotes learning and is therefore commonly used is tutorials.

- *Alphabetical.* This is the common—probably the only appropriate—approach for organizing reference information.

- *Psychological.* If you follow this approach, you would place the most important information in the most strategic place. Depending on the situation, that place can be anywhere, not necessarily the beginning.

- *General to specific.* Sometimes you may want to begin with, say, a hypothesis and then move on to supporting information.

- *Problem-solution.* You may want to use this approach in providing support information for error messages. First you describe what went wrong, then you explain how to correct it.

Unlike the print medium, the online medium allows multiple paths to topics. Some important online navigating techniques are described here. See Horton [1] for a comprehensive coverage.

- *Sequence.* Prescribed path of screen after screen (or page after page) of information. There are some variations to the sequence approach. A *rigid sequence* has a one-way flow and is quite restrictive for most situations. A *sequence with alternatives* approach allows alternative paths to information while retaining an overall sequential flow. A *sequence with backtracking* approach allows users to repeat certain sequences, whereas a *sequence with shortcuts* allows users to skip a series of topics. You can also consider a *looping sequence,* which is a rigid sequence that repeats continually.

- *Hierarchy.* This is a composition of topics, subtopics, sub-subtopics, and so on. Users start at the highest level generality and go on to lower level specifics. A hierarchical flow need not be this rigid. You can provide shortcuts, allowing users to jump directly to another relevant topic.

- *Web.* This is a structure that consists of topics linked for the ultimate in user exploration. The web structure is common in hypertext systems. In a *rigid web*, every topic is directly linked to every other topic. The number of links increases dramatically with the number of topics. Maintaining such large numbers of links means more storage space and slower systems. In a *partial web*, you will limit connections and link topics to only a few other topics. Unlike a rigid web, a partial web is less powerful but more predictable. Keep in mind that predictability can seem restrictive to experts, whereas full exploratory power can be bewildering to people who are new to the software.

13.2.3   Structured presentation

Structured presentation techniques make it easier to find information that would otherwise be "hidden" in a prose form of writing. In the following paragraphs, we will see some of these techniques. For details see Brockmann [2].

Matrix structure   Consider the matrix structure if you want to describe cause and effect, and if users know what to look for. See Figure 13.1.

Problem	Cause	Solution
Fonts do not print as smoothly rounded characters.	Bitmapped fonts designed for DMPs do not print smoothly.	Choose another font such as TrueType font.
Large gaps appear between text.	Video driver may not be VGA.	Install a VGA driver if not already installed.

Figure 13.1 Example matrix structure.

STOP Shown in Figure 13.2, STOP is an acronym for *sequential thematic organization of proposals*. In STOP, all information is broken down into topics, each placed on a double-page spread. The left page has a headline, thesis statement, and text of about 500 words. The right page has supporting graphics.

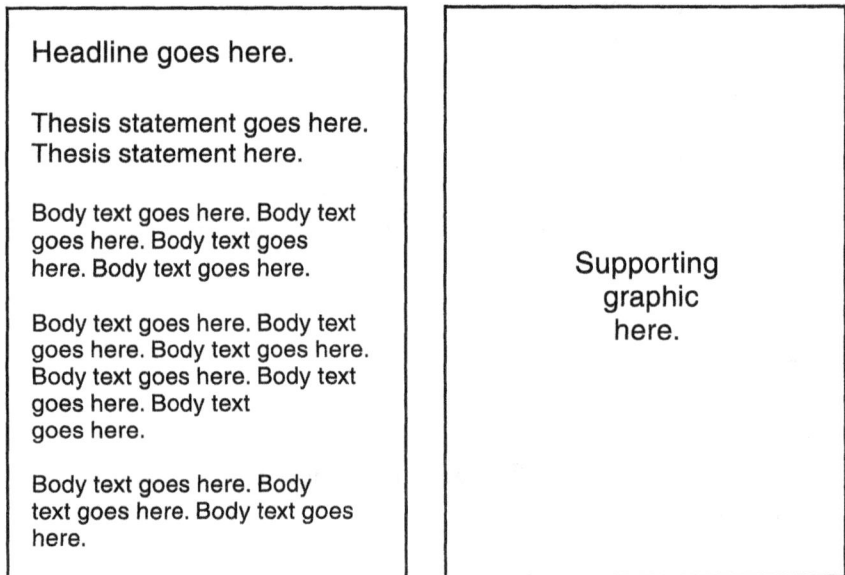

Headline goes here. Thesis statement goes here. Thesis statement here. Body text goes here. Body text goes here. Body text goes here. Body text goes here. Body text goes here. Body text goes here. Body text goes here. Body text goes here. Body text goes here. Body text goes here. Body text goes here. Body text goes here. Body text goes here.	Supporting graphic here.

Figure 13.2 STOP.

The headline should summarize the topic. The thesis statement should overview the main idea of the two pages. The user is given the opportunity to read only the statements written on a generalized level because the thesis statement, which is on a general level, has been visually set apart by format.

There must always be a supporting graphic. Some companies place examples here instead. The graphic or example should have an expanded caption of three to four lines of text.

The advantage is that by looking at the graphic and reading the caption, users can receive much of the information on the topic. STOP has at least two disadvantages. As each topic is only two pages long, the hierarchy of topics in a manual can be obscured. Moreover, the requirement for a graphic on the right page can lead to a use of graphics and examples that are unnecessary from the users' viewpoint.

Playscript The playscript presentation technique has a two-column format that helps users quickly scan the information. Concisely and effectively, it answers the questions, who does what *when*. "Who" are the user roles (or actors) appearing in the first column, "what" are the user tasks (or actions), and "when" are the numbers indicating the sequence.

Structured writing This technique borrows the two-page format from STOP and uses the playscript technique for presenting procedures (see Figure 13.3). Structured writing makes it easy to scan. It also helps the writer stick to the point.

The page layout is a two-column format. The first column has the headings and the second column has the corresponding textual information. User attention is drawn to the headings because of their left-side location and the large white space.

Textual information is presented in visible blocks. Each block localizes all the information on one topic or one aspect of that topic. Horizontal rules separate the topics. A thesis statement appears at the top of the page. An overview block appears at the top of the page or chapter. It relates new ideas to previously discussed concepts.

In structured writing, sentences are written in similar style wherever possible, and lists are frequently used.

Adding an employee	
What is this task	You can add a new employee by specifying certain required details such as name and date of joining.
When to perform	You should enter the details whenever a new employee reports at a location, on appointment.
Prerequisites	You can only do this if an appointment order was earlier generated for the same person.
How to perform	To add an employee: 1. From the main menu, select Employee. 2. Enter the new employee details. 3. Press Enter.
System response	Once you specify all the required details, the system assigns and displays a unique employee number.

Figure 13.3 Structured writing.

13.2.4 Headings

The purpose of a heading is not just to mark the beginning of another section, but to identify the content in order to improve retrievability and save time wasted reading wrong sections. Headings like "Introduction," "Overview," and "Appendix" say little about the content of sections. Rather than a heading like "Appendix A," explain what the appendix contains with a heading such as "Appendix A: Internet basics." Besides improving retrievability, good headings make the page appear more inviting and help the writer stay on the topic.

Convey to the user the contents of the part, chapter, or section the heading identifies. Whenever possible, the heading should be understandable out of context. Headings for task information should be task-oriented. For this, start the headings with gerunds, as in "Performing simulation." This heading would be even more useful if you included the purpose of the task, as in "Performing simulation to project revenue and profit."

Levels of headings, when differentiated via good formatting, give the visual impression of major and minor topics, and their relation to one another. Techniques to differentiate levels include different position, typeface, type size, and style.

You can also use these guidelines to write your headers, footers, figure captions, and table captions.

13.2.5 Roadmaps

In printed information elements, you can use the preface and the chapter introduction to provide useful roadmaps for users.

Preface The preface is about the information element. State the purpose of the element. Will it help users know the software's features or code application programs? Mention user groups. Indicate more than just the user's job title: "The manual is intended for system programmers who use the assembler language to code programs." Mention the prerequisites for using the software system in terms of skills and knowledge. Briefly describe the content and organization of the printed element. Also mention the conventions and terminology used. List other support information elements users may have to read. Finally, mention any useful how-to-use tips.

Chapter introduction The chapter introduction is the first few paragraphs of a chapter. In a chapter that describes a task, the chapter introduction should contain a quick definition of the task, the target users, and information such as when to perform the task.

13.2.6   Table of contents

The table of contents (TOC) typically shows two to four levels of headings. If the number of chapters or sections is very large, you might want to first provide a short table of contents, followed by a detailed one. If you have organized the information element into parts, you might want to first provide a short table of contents, followed by a partial TOC at the beginning of each part. Partial TOCs usually appear in divider pages. They should not be over a page.

The online table of contents is similar to that in printed elements. In hypertext Help systems, the TOC often only contains a list of higher level topics. When users click on a heading, lower level headings appear.

If you have a significant number of figures and tables in your printed element, you might also want to provide a "List of figures and tables." If this is a long list, you can have separate lists for figures and tables.

13.2.7   Index

The index is a list of entries (or topics) with useful page references. Create an index for all printed elements over 25 pages. Consider providing a master index if you have too many printed elements. You should also plan to provide an index for online elements.

How do you ensure the completeness of an index? Anticipate the entries that users will look for. Include tasks, labels, concepts, abbreviations, and acronyms. Include all topics referred to in the text. If some information covered in your information element does not have a word found in the index of a competitor's manual, add that word to your index as well. Use the guideline of one index entry for every 100 words of text. Expect more for reference manuals and much less for other types of manuals. Note that a large index in a printed element need not be necessarily better because it could mean longer search time.

Creating an index can be tedious. Even book publishers often seek external agencies that specialize in indexing. Your word processor can help in picking up page numbers, updating page numbers, organizing topics, and formatting index pages. But you still need to first identify the entries that should go in. For this, look through the information element for entries. List all useful entries. Break large topics into secondary entries. Word them from the users' viewpoint and language. List all possible combinations of each entry. Here is an example:

    editing
        files 8
        macros 24
        ⋮

    macros
        defining 7
        editing 24

Users want to know what something is or how to do something—not *where* words or phrases appear in an information element. Identify and include only page references that will be useful. A guideline is a maximum of four references to an entry. To reduce the number of page references, you can try creating subentries as shown here:

    Web server
        detailed description 22, 35, 42, 49
        general description 20

Support cross-referencing via "See" and "See also." For style-related issues, use a style guide such as the *Chicago Manual of Style* [3].

The online index is similar to the index in a printed element. Subentries may appear indented just as in a print-based index. Users can select any entry to display the relevant information. To access the required entry, you can allow users to select the first letter of the entry so that you can display all entries that begin with that letter. If required, users can scroll and select the required entry.

13.2.8   Hypertext links

Hypertext is one of the best things to happen to online information. Whereas paper imposes a linear organization on your information element (one topic following another), online hypertext allows information to be organized as a "web" of separate topics. Each topic in the "web" can be linked to any or every other topic. See Figure 13.4.

Applying a style

Once you've created a **style,** you can apply it to objects in your drawing with the Object menu. Styles in the default styles **template** are applied the same way.

You can also use keyboard shortcuts to apply Paragraph text styles. The Styles window lists the shortcuts for the current template. It also provides a command for changing and adding assignments.

A group of **styles** stored in files with a TMF extension.

You can open a template and apply its styles to any illustration.

The hyperlinked word or phrase is highlighted.

The pointer users can use to select (click) a hyperlinked word or phrase.

Information displayed when users click on Template.

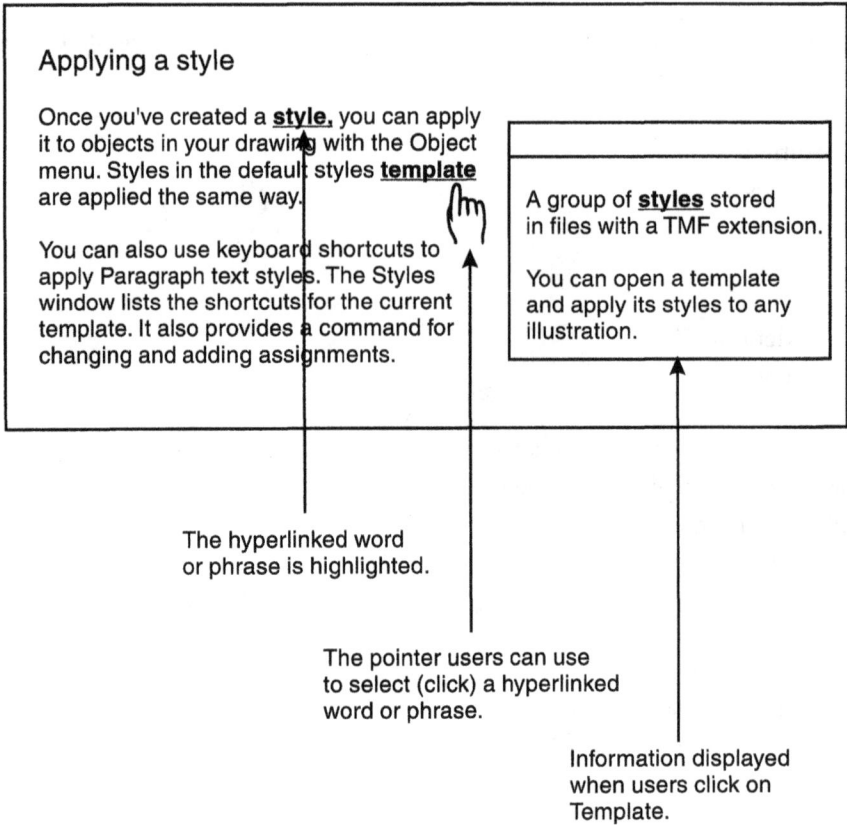

Figure 13.4   Hypertext retrieval.

Hypertext is advantageous to users. It very significantly improves online retrievability. It helps users avoid the process of going through a TOC or index, which can take relatively more time. If designed well, hypertext can support user exploration by providing multiple paths to information. It can also provide shortcuts through information, taking users instantly to the required information.

13.2.9   Online search

A simple online search is similar to the Find command in word processors. Users enter a word or phrase in the Search field. You display either the requested information or a list of related topics from which users can select the most appropriate topic. See Figure 13.5.

You may want to provide advanced search, where users search by using operators and expressions. An operator is a symbol such as an asterisk (*) that controls the search. An expression is any combination of characters and operators that specifies a pattern. Here is an example of an expression:

b?t

This expression finds three-letter words beginning with "b" and ending with "t."

In online query, users request topics by describing the characters of the information required. You display topics that match the characters. For example, the query (SUBJECT="usability engineering") AND (PUBDATE1996) will list all topics on usability engineering published after 1996. One disadvantage with this approach is that users may have to learn the query language.

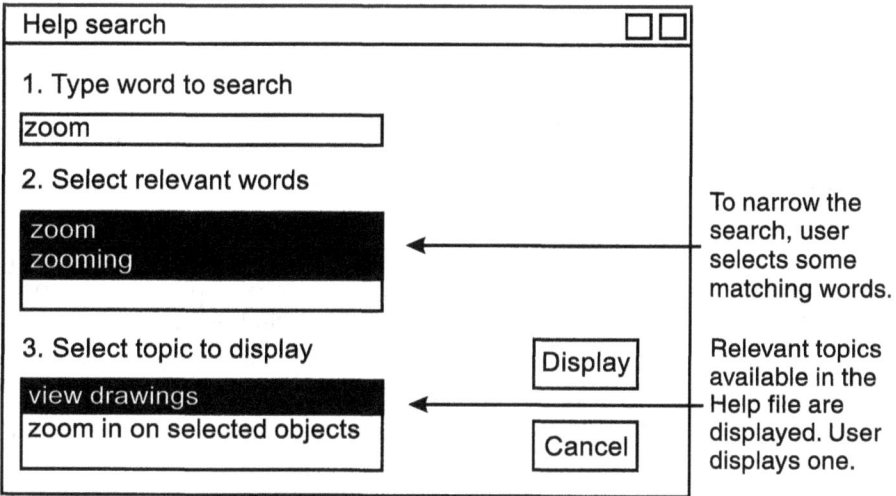

Figure 13.5   Online search.

13.2.10   Online navigation

Many facilities exist to make online navigation easy. Consider facilities such as Previous, Next, Backtrack, Home, scrollbar, page (or screen) numbers, and shortcuts. The Previous button displays the previous screen. The Next button displays the subsequent screen. Backtrack keeps track of topics displayed earlier so that the user can redisplay topics in reverse sequence. The Home button displays the top screen in the menu hierarchy. Scrollbar and paging help users in moving within a topic. Paging is preferred so that you have control over the look of a "page" of information. (This, however, will only be useful if the topic preferably ends within a page, not when a sentence or paragraph runs over to the next page.) Shortcuts help users tunnel through layers and go directly to specific facts.

You can also provide Index and Search buttons to help users go to the index and search facilities.

Bookmarks are online equivalents of dog-ears. Topics users want to return to are "bookmarked." Users can now go exactly where they want. Finally, hypertext can help users jump to online cross-references.

13.2.11   Cross-references

Cross-references are not generally liked by users, especially in printed elements. Cross-references may be required if related information is presented in different places. Organize your information element such that you can avoid references to other pages or other information elements. It may also be better to duplicate information (if it is, say, a few paragraphs) than to give cross-references.

There can be many types of cross-referencing. In *directional reference*, you will use a word such as "above" or "below." For example, "These concepts are described below." Just make sure that "below" is neither two pages down nor actually *above* in a two-column format. When you make an *out-of-document reference*, you should give the exact title of the printed element in italics. For example, "For more detailed information about job control language, see *Job Control Language User's Guide.*"

Use terse cross-references like "See Part I" only if the reason for the cross-reference is obvious from the context. When making a cross-reference, tell users why you are referring them to another location or mention the kind of information they will find. For example, "To find out how to forecast resource usage, see…." Use heading, head number, and page number to indicate exactly where the user should go. For example, you can write: "See Chapter 7, *Monitoring the server*, on page 77." However, you may want to avoid page numbers if your information development tool cannot track and change the references as the information element changes.

13.2.12   Tables

Use tables to give quick reference information. Instead of "Press F1 to get help. To scroll up you must press F7 and to scroll down you must press F8," put it in a table format as shown here:

Press This	To
F1	Get help
F7	Scroll up
F8	Scroll down

The advantages of a table are many. Well-designed tables show relationships clearly, emphasizing similarities and differences. You can group similar items (or treat them in the same way) and separate dissimilar items. Information in a good table is concise, essential information. Retrievability is improved and users quickly get the big picture.

However, tables lack the contextual clues that textual information can provide. Because they lack context, tables should be clear and self-contained. Users should never have to search in the text to find what the table means. Another disadvantage is that some users will need explanatory material, which cannot be provided in table format.

Title and number all tables. Every table should be referenced in text. Place tables after, and close to, the reference.

13.2.13   Numbering

You can number headings, pages, and lists. Numbering helps in retrievability and cross-referencing.

Numbering the headings Typically, headings in printed elements are numbered. Do not overnumber, such as the 3.3.12.3.2.2.10 found in a Defense manual. Try and limit to four levels. Head numbering may also be required in an online manual, especially if it is a duplicate of a printed element.

Numbering the pages (screens) Provide page numbers for printed elements of all sizes. Decide what numbering style would be best considering the size of the information element and the way it would be used. For online elements, consider numbering as in: "Screen 2 of 4."

Numbering the lists Number procedural steps. Also number things for which a count is required and has been mentioned. For example, if you say "The package comes in the following six diskettes" then number the list of diskettes.

13.2.14   Visual and physical devices

Some visual devices are graphic images and highlighting techniques. Graphic images can be pictorial roadmaps and icons like revision symbols. Highlighting, if done carefully, can attract attention to the required items on the page or screen. For highlighting, consider using typography, color, blinking, reverse video, and layout. Take care not to invite attention to unimportant items such as borders.

An example of physical retrieval devices is the set of divider pages you can use in large printed elements. Use them to show the major divisions of a printed element. Include a partial table of contents on each divider. You might want to have these dividers in a different color to catch attention. Or, you can use tabs for the same purpose. Dividers can be hard or regular pages. Hard dividers with tabs can be significantly more expensive to produce.

13.3   Readability

Often, you can identify an effective document with only a glance. An ineffective document contains "wall-to-wall" text ... bulky paragraphs

... long sentences that leave the audience breathless, monotonous use of type... [4].

Effective communication depends not only on content, but also on the way the content is presented. Hard-to-read information can reduce the user's efficiency. Worse, it may not be used at all. Compared to paper-based text, online screen text typically has poorer resolution. More often, the screen is not carefully *designed* for easy reading. These are reasons why screen reading is said to be about 25% slower, and also less accurate.

Improved readability invites users to read, and speeds up reading. Readability attributes like type size also affect retrievability and pleasantness.

Readability can depend on certain things over which you do not have control. This includes screen resolution, user's visual acuity, reading distance, and how the user "uses" the information. Readability can also largely depend on things that you can design. This includes page layout, typography, and color. These are described in the following paragraphs.

13.3.1 Layout

The various layout attributes of a page (or screen) impact readability. They include columns, white space, margins, text line length, line spacing, and text alignment.

Design for a proper balance of text, graphics, and white space. In general, more *white space* means better readability. However, too much white space can make online elements unusable and printed elements bulky. A two-column page design where column one has the headings can be useful, especially in printed elements. Have frequent headings, at least one in every page. To separate certain items, consider boxes and lines.

Long lines of text are harder to read. Limit *line length* to around 12 words. For online elements, give default line length, but let users have the option to customize. Make *line spacing* at least 1 / 30 of the line length. Too much will destroy the integrity of paragraphs.

Control *text density*. Do not pack text tightly. Keep in mind that English text expands when translated. When it comes to *text alignment*, go for left alignment of text wherever possible. Justified text reduces reading speed by 10%.

13.3.2   Typography

Consider differentiating heading levels, body text, and other page (or screen) items using different *typefaces*. But do not use too many. Two should be just fine in most cases. *Type size* should be large enough to clearly see each character and small enough to see a set of words at a glance. Online type size is usually described in terms of visual angle. For a display viewed at 20 inches, a 0.35-degree visual angle requires a character size of about 0.12 inch. At the screen resolution of 75 dots per inch (dpi), each character will be nine pixels high. A visual angle of 0.35 degrees is recommended for continuous reading. For scanning, visual angle should be at least 0.5 degrees.

Capital letters are harder to read. Lower-case letters are preferable, which give a word a distinctive shape.

As far as *letter spacing* is concerned, it is preferable to use proportional rather than fixed spacing.

Legibility improves with increased *contrast* between text and background. Increased contrast can compensate for small type size. A black-and-white combination has the greatest contrast.

13.3.3   Color

You can use color more than just as a device to "add color" to information elements. Use color for highlighting portions of text for improved readability and retrievability. Color, of course, impacts pleasantness, too. But, what is pleasing for some users may be distracting for others. Therefore, while you provide default colors, give users the option to change them according to their taste.

13.4   Clarity

Accuracy aside, clarity is perhaps the single most important quality objective of information. Clear information is that which is easily, quickly, and correctly understood by users. See the list below for various ways to achieve information clarity.

- User viewpoint writing;

- Contextuality;

- Conciseness;

- Coherence;

- Consistency;

- Distinctness;

- Examples;

- Illustrations;

- Listing.

13.4.1   User viewpoint writing

Members of the software engineering team are familiar with the design and working of the software. Users generally are not. Their knowledge or information requirements are therefore different. Moreover, their characteristics in terms of education and job experience are probably different as well. Therefore, if you want to communicate with these people, you need to start where your users are.

Provide all the information target users need. Do not assume that something is self-explanatory or known to users. Omitting major topics can cause serious delay until a user can get information from other sources. Omitting minor topics can cause temporary delay or force a user to use trial and error.

Use computer jargon with discretion. In general, computer jargon and abbreviations are not clear to people outside the computing profession. They may be appropriate to use if the target users will understand. For example, a doctor while speaking to another medical professional can say, "A myocardial infarction is contraindicated." However, when they speak to nonmedical people, they would just say "No heart attack." Similarly, you should adjust your vocabulary to suit those you "speak" with. If you decide to use jargon, define when it appears first. Also, repeat the definition in the glossary and, if possible, as a footnote. Consider this headline: "Customs slaps duties on OTP EPROMs." If users understand *OTP EPROMs*, how much better it is than having to say, "one-time programmable erasable programmable read-only memory chips!"

You must give special attention to the user-viewpoint writing of messages, which appear while users are in the middle of task. A technically accurate message can be as useless as no message at all if it is not conceived and written from the user's point of view. For example, a message such as "Query caused no records to be retrieved" is useless unless the users are programmers. If you are writing for nonprogrammers, you are better off avoiding this kind of language: record, row, column, table, fetch, mapping, processing, writing on the database.

Talk directly to users. Avoid using the word "user." For example, instead of "The user should press ENTER" you would say "Press ENTER."

13.4.2 Contextuality

Context sensitivity obviously belongs to the online medium. Context-sensitive information matches the user's situation and is therefore immediately useful. The software system (even automatically) displays such information based on, for example, the user's most recent command, selection, or error.

13.4.3 Conciseness

Conciseness is saying it in as few words as possible, without impeding clarity. The sentence "You can modify all system configuration parameter values" is brief, but not very clear. In contrast, the sentence "You can modify the values of all parameters that control the configuration of the system" is a bit longer, but clearer. It uses all the words needed to get the message across, but no more. You can achieve conciseness when you follow the techniques described in the following paragraphs.

Simple writing Big words or literary phrases will not enhance your writing or your image. Often, they do just the opposite. Simple writing is appropriate for business. It is also easy to translate.

Use what are called first-degree words: "face" instead of "visage" and "stay" instead of "reside." First-degree words immediately bring an image to the user's mind while others must be "translated" through the first-degree word before you see the image. Replace verbose phrases like "at the present time" and "in the event of" with "now" and "if," respectively. Avoid Latin abbreviations such as "etc." Instead of "C, C++, etc." you can say "C, C++, or any other high-level programming language."

Keep things short   Here are some rules of thumb. Use short words, that is, one- or two-syllable words. Try and limit sentence length to about 20 words and the number of sentences in a paragraph to around four.

Cut down words by avoiding grammatically complete sentences wherever appropriate. Consider following this guideline when you design labels that have to be fit into very little space on the screen and when you write procedural steps in support information elements. In labels, for example, prefer "Export file" to "Export the file."

Avoid redundancy such as in "up to a maximum of." Finally, when you've finished, just stop.

Imperative tone, active voice   Instead of "Individual project files must be consolidated into a single window," say "Consolidate individual project files into a single window." This is likely to result in shorter, clear sentences.

13.4.4   Coherence

Make your writing cohesive so that users advance smoothly, not having to jump forward and back to make connections. Let users know up front how the information element connects with their interests. Give background information if useful.

Paragraphing is a method you should use to divide information into paragraphs to make writing more readable and more clearly understood. Have a topic sentence to convey the central idea of each paragraph.

Consider this paragraph:

The most common error recovery approach is "crash and burn." The other is the "hang on until the bitter end" approach.

Instead, try something like the following, where the first sentence is the topic sentence:

Two error recovery approaches prevail in mainframe operating systems. The most common is the "crash and burn" approach. The other is the "hang on until the bitter end" approach.

The topic sentence need not always be the first sentence in the paragraph. Sometimes it is even the last. Sometimes, it is just implied.

Check for missing transitions. Transition helps in the smooth flow of thought from sentence to sentence and from paragraph to paragraph. To indicate that one subject is over and that another is beginning, use transitional words, phrases, sentences, paragraphs, or headings. Example transitional words are "and," "but," and "for."

Check word order. For example, "The conference will accept technical papers from engineers of any length" should be "The conference will accept technical papers of any length from engineers."

Ensure parallel construction. The sentence "I studied electronics and how to do programming" is not parallel. Instead, you say, "I studied electronics and programming." Also ensure parallelism in lists by beginning every list item with similar types of words or phrases.

13.4.5   Consistency

I have seen a software name used in four different ways in the same manual. This confuses users. Look for consistency in terminology, style, tense, person, tone, singular-plural, conventions, format, and word choice. Users need to feel confident that the conventions and highlighting devices they encounter across the information elements always mean the same thing. Inconsistent use causes confusion, heightening user frustration and lowers their opinion about your company and your products.

For users, consistency is getting a dialog box labeled "Convert Text to Table" when they select the option labeled "Convert Text to Table." Consistency is also parallel structure of a group of options.

13.4.6   Distinctiveness

Distinctiveness, especially in labels, ensures that users will not confuse one label with another. Avoid subtle distinctions, such as between "View" and "Show." Users may waste time selecting the unintended option.

Distinctiveness is especially critical in command labels, where incorrect entry may produce unintended commands.

13.4.7 Examples

Good examples can clarify things. The sentence "In some languages, expressions are evaluated from right to left" may not be clear to some users. Consider providing an example as shown below:

> In some languages, expressions are evaluated from right to left. For example, in (8-5-2), first (5-2) is evaluated. It yields 3, which replaces (5-2). Finally, (8-3) is evaluated, resulting in 5.

Good examples can often effectively replace lengthy paragraphs of explanatory text. See how the example in the following instruction helps.

> If you do not remember the full name of the company, just type the first few characters, For example, if you want to inquire on Dun & Bradstreet Corporation, just type "dun."

Usually one example is fine. In fact, more than one can confuse users. So unless you want to reinforce, demonstrate, or clarify different facets of the same concept, one good example is enough. Here are some types you might want to consider: concept or procedure, scenario, sample, case, model, metaphor or analogy, and "running."

Make sure that examples are clear; otherwise, they defeat their own purpose. Examples must also be realistic—that is, they must be likely situations. Ensure that they are accurate and will execute successfully. Place the example near the text it clarifies.

13.4.8 Illustrations

Compared to text, good illustrations can often communicate information faster and more clearly. Good illustrations also make information elements inviting and friendlier, and reduce the amount of text required. Consider an illustration to reduce text, summarize concepts, show how something works, or to aid recall.

Illustrations should not be cluttered with lots of details. Do not try to portray many things in one illustration. It is best to communicate one idea per illustration. For each illustration, provide a title and a number. Every illustration should be referenced in the text. Write something like "Figure 1 shows..." or "See Figure 2 for..." Place illustrations after, and close to, the reference in the text. Use descriptive words in the caption for illustrations: "Table 1. Information requirements for novice users." Consider providing a "list of figures."

Icons should be designed to display well on the screen. You should also consider international users while designing icons.

Make sure that text appearing in illustrations is readable. Also, be sure that the elements of an illustration (lines, text, etc.) are balanced. Make line weights heavy enough for easy viewing and pleasing appearance. Make sure that any photographs you reproduce are clear and easy to interpret. They should be in focus and have enough contrast. Diminish or remove irrelevant elements from photos.

13.4.9 Listing

Listings pick prose buried in paragraphs and highlight them concisely. Consider this sentence:

> By sharing resources between projects, you can add shared resources to a new project quickly, review resource usage and costs across projects, identify overallocated resources and correct the cause of the overallocation, and print resource reports that provide information about resource usage across projects.

Now see how listing improves its clarity:

By sharing resources between projects, you can:

- Add shared resources to a new project quickly.
- Review resource usage and costs across projects.
- Identify overallocated resources and correct the cause of overallocation.

- Print resource reports that provide information about resource usage across projects.

That is a bullet (or unordered) list. The following is an example of a numbered (or ordered) list:

To enable remote file manipulation:

1. Choose Remote File Manipulation.
2. Choose a resource.
3. Click the Yes radio button to enable remote file manipulation.
4. Click OK.

Standardize on style in terms of case of first letter, punctuation, and line spacing. Avoid overuse of lists.

13.5   Usability

The information design goal "usable" applies to printed elements in terms of handiness. Bulky or heavy manuals can be hard to use. The things you should control are page count, size, shape, cover and paper quality, and binding.

The "usable" goal is largely an issue of the online media, where users are likely to use the information while they perform tasks. Keep in mind the following requirements while designing online information elements:

- Do not completely hide the application screen on which the user was working.

- Do not completely remove information when users try to apply that information on the application screen.

- Ensure that users can quickly and easily remove the information.

- Ensure that the user interface for Help is consistent with the rest of the software.

13.6 Pleasantness

The use of color and verbal tone impacts pleasantness. You can use color effectively for enhancing retrievability and readability as well as pleasantness. As far as tone is concerned, strive to be polite and helpful while avoiding excessive praise.

13.7 Accuracy

Technical precision   If information is not accurate, achieving all the other design goals discussed in this chapter is a waste. Inaccurate information can be harmful. Have a technical expert inspect every information element. But that is not enough. Information, especially examples, must be tested for accuracy. You should also have a change control mechanism that ensures that all software changes are reflected in information.

Look for hidden contradictions, such as in this example: "The file requires about 205 bytes, including 135 for abc and 65 for xyz." Check all references: section numbers, page numbers, titles, cross-references, and index.

Language precision   Use correct grammar, spelling, and punctuation. Inaccuracies can mislead users. Use the word "can" to indicate capability, "might" for possibility, and "may" for ambiguity. The word "that" identifies, as in "Tabby has two cars. This is the Toyota that is two years old." The word "which" amplifies as in "Tabby has two cars. This is the Toyota, which is two years old." Before you write, decide which is accurate: "must" or "should"? "Is" or "is being"?

References

[1] Horton, William K., *Designing & Writing Online Documentation: Help Files to Hypertext*, New York NY: John Wiley & Sons, Inc.

[2] Brockmann, John R., *Writing Better Computer User Documentation: From Paper to Online,* New York, NY: John Wiley & Sons, 1992.

[3] The University of Chicago, *The Chicago Manual of Style*, 13th Edition, Chicago, IL: The University Press, 1982.

[4] Mancuso, C. J., *Mastering Technical Writing*, Reading, MA: Addison-Wesley Publishing Company, 1990.

14

Evaluating Information: Two-Level Approach

P LANNING for information quality involves setting goals and deciding how you will evaluate to see if those goals have been met. Whereas this chapter is about evaluation, goal setting is explained in Chapter 6.

Information is good quality only when it contributes to improved software usability. When your ultimate goal is improved software usability, the traditional document-centric evaluation effort will not take you far. You need an approach that will improve individual information elements so that they—in combination—achieve improved software usability. In addition, your approach should also cover interaction information—that is, labels and messages. This chapter describes a two-level evaluation approach to help achieve improved software usability. You will learn how to use common testing, review, and editing methods to perform the two levels of evaluation.

The two-level evaluation approach (Figure 14.1) helps improve your individual information elements so that they will together, and with the interaction components, maximize software usability. The first level, called *integration evaluation*, involves the testing of the entire software system, including all the software usability components. The second level, called *CDG evaluation*, involves testing and reviewing each information element to determine to what extent it achieved its CDGs.

14.1   Level 1: integration evaluation

The importance of an integrated validation test cannot be overstated. Since components are often developed in relative isolation from each

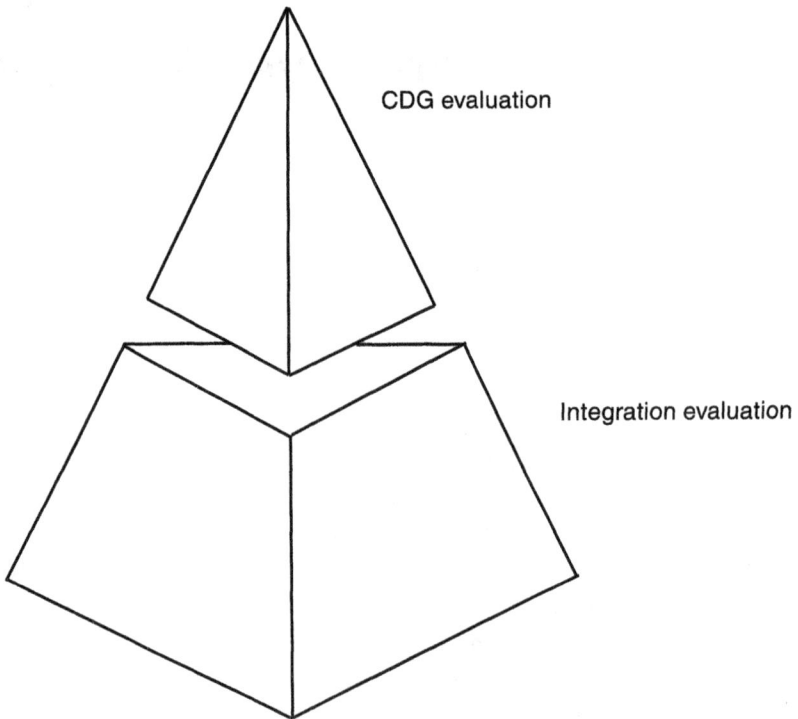

Figure 14.1   Two-level evaluation.

other, it is not unusual that they do not work well together. It behooves an organization to discover this prior to release since, from the user's viewpoint, it is all one product and it is expected to perform that way [1].

Integration evaluation involves the entire software system, covering all four software usability components, including the information components of labels, messages, online support elements, and printed support elements. Software usability components are—as one piece—tested by users for usability. Although it is called Level 1 evaluation, integration evaluation is not the first you perform. Actually, you will first review and edit an information element before you evaluate it as part of the integrated software system.

There are two major reasons why you need integration evaluation.

Integration is one reason. Integration evaluation is required to see whether or not the individual information elements are effective as an integrated whole. This evaluation identifies and helps remove issues related to information as it is used while performing tasks. Integration evaluation helps you confirm your information architecture and provides answers to questions such as the following:

- What types of information are users looking for: definition, reason why, how to, example, figure?

- Which information elements do users use? When? And why?

- What kinds of users use which information elements? Why? In which situations?

- What types of information do users want online versus in print?

- Which information should be repeated in a printed manual?

- Should the question "How to…" be answered in the message text or in a separate message Help?

Task context is another reason why you need integration evaluation. Suppose you want to evaluate the quality of labels. One evaluation method is to show labels to representative users and ask them what each label means. This way you know if users understand the meaning. However, because the context of label use is absent, you wouldn't know about all the aspects concerning label use—that is, noticing the label, reading it, understanding its meaning, and acting correctly. Context of task performance is therefore essential for evaluating labels and other information elements.

At least two rounds of integration evaluation are required, one when the first product prototype is available and another prior to shipping of the software system. The evaluation method typically used is usability testing in a laboratory setup.

14.2　Level 2: CDG evaluation

CDG evaluation involves the testing, reviewing, and editing of each information element. You should perform this type of evaluation to determine to what extent the CDGs you defined for each information element have been achieved. CDGs are information design goals such as retrievability and distinctness, which are critical for the success of an information element. CDG evaluation keeps your testers, reviewers, and editors focused on these goals so that the resulting information elements help users complete their tasks easily.

Right from the prototype stage and until the information element is fully developed, you could perform any number of CDG evaluations. Some, such as validation testing, are for determining the accuracy of information. Others are for evaluating, for example, the effectiveness of retrievability devices or the understandability of labels.

14.3　Evaluation process

Plan to evaluate how well your information elements contribute to improved software usability. Specifically, plan to evaluate how well the information elements meet the information quality goals you have earlier defined. The technical writer/manager should plan in consultation with the usability engineer and project manager. Record the planning details in the UCID plan.

The evaluation process consists of these activities:

1. State evaluation objectives.

2. Decide on evaluation methods.

3. Perform evaluation.

State evaluation objectives In Chapter 6, we have seen that CDGs are general goals and that, based on them, we need to come up with more specific, measurable goals—called *information quality goals*. You should have separate sets of information quality goals for each information element. From the goals, it should be clear what user performance data (or metrics) and user preference data you need to collect.

Define evaluation objectives based on information quality goals. Evaluation objectives can be the same as the software usability and information quanity goals you defined earlier. An example of an evaluation objective could be to compare different online retrieval techniques to learn which one helps users find information more quickly.

Decide on evaluation methods Based on evaluation objectives defined earlier, you must plan what methods you will use to evaluate each UCID deliverable. Evaluation methods could be different for each one. In the UCID plan, list all the evaluation methods you will use. Also describe the error management procedures you will use to ensure that all accepted findings and comments are implemented (see Chapter 4).

There are many evaluation methods to help you determine if goals are being met. For integration evaluation, a method such as a lab-based software usability test is required. You probably already have this method as part of your organization's user-centered design process. For CDG evaluation, you will need validation tests, reviews, and editing. In addition, you can consider other methods such as interviews and questionnaires.

Performing evaluation Before you can perform any evaluation, you need to prepare for it. Preparation includes activities such as designing tests, briefing reviewers, and preparing test scenarios. Once the actual tests, reviews, and editing are over, you need to perform postevaluation activities. This can include the preparation of detailed reports, as in the case of a lab-based software usability test.

14.4 Testing information

One test in which all your information elements must participate is the product usability test. The type and number of usability tests depend on the user-centered design process you work in. More likely, there will be three usability tests: the first with the product prototype, the second when a substantial product has been coded, and the third at preshipment. Each is an integration evaluation from the UCID viewpoint.

In addition, you will have second-level tests (CDG evaluation) that focus on individual information elements. For example, you could test messages along with corresponding message Help to ensure context-sensitivity and completeness. Or, you might want to have a test to validate the accuracy of the examples used in various information elements.

In this section, we will see how a lab software usability test is performed. Most of the techniques described, however, can be used for your CDG evaluation as well. Later in the section, we will also discuss validation testing which is a CDG evaluation.

14.4.1 Lab software usability testing

Here is an overview of a typical laboratory test that can be used for integration evaluation. From an information design viewpoint, the objective of this test is to see how well the various information elements work together *and* with the other software usability components. In a lab test, test participants perform tasks with the software system. Typically, a usability engineer leads an observation team consisting of programmers and technical writers. Software tools and other devices are used to gather quantitative data on the use of the software system. The test is often video recorded. Once the test is complete, the observers get subjective details from the test participants. All the quantitative as well as subjective data collected at the end of the test via interviews is analyzed to find usability problems. Then, the observers recommend solutions. Here is the list of important activities:

■ Define test (evaluation) objectives;

■ Prepare test materials;

■ Get people;

■ Set up lab;

■ Conduct test;

■ Debrief test participants;

■ Analyze data and recommend solutions.

14.4.1.1　Define test (evaluation) objectives

Test objectives are usually the same as the software usability goals and information quality goals you have defined. Test objectives should be clear, concise, and measurable (or observable). The design of the test and the data to be collected both depend on the test objectives. Consider what ones will address the test objectives and how you will collect, record, and analyze them. Data you collect can be quantitative user performance metrics (e.g., number of times Help is accessed) or subjective user preference measures (e.g., overall ease of use).

14.4.1.2　Prepare test materials

Depending on how the test is designed, you may have to prepare materials, including the following:

Screening questionnaire　Give this questionnaire to prospective test participants to check if they are potential users of the software system. Get background information about the participants: native language, job title, job role and responsibilities, education, computer exposure, experience with similar software, and age. Include questions such as, "Do you use a computer?" and "Of the total time spent using a computer, what percentage do you spend using a word-processor?"

Agreement forms　Design a nondisclosure agreement form that test participants will sign before taking the test. This requires that participants do not share information about the software system with the outside world.

Also design a video consent form. By signing this form, participants agree to be videotaped.

Task scenarios   A scenario is a task within a real-world-like situation that participants will perform using the software system. While writing scenarios, describe the end results participants should achieve and the reasons for performing the task.

Post-test questionnaire   Data such as opinions and suggestions that cannot be directly observed are captured via a questionnaire administered after the test. An example: whereas you should find through direct observation how often users access the index, users' feelings of your index's retrievability should be captured in the post-test questionnaire.

Consider providing a Likert scale, where users indicate their agreement or disagreement:

Labels are meaningful (circle one):

Strongly disagree
Disagree
Neither agree nor disagree
Agree
Strongly agree

You can have checkbox questions:

Check the item that best describes your feelings:
❏ I always start with a tutorial
❏ I only use a tutorial when asked to
❏ I never use tutorials

Or you can have fill-in questions:

I found the tutorial useful because

_____

_____

14.4.1.3   Get people
Identify the following people and make sure that each person understands his or her role in the test.

Test participants   Identify test participants, who will test the software system. They should be actual users or people with the characteristics of actual users as described in the user profiles. Decide on the number of users who will test. Contact and screen potential participants well before the test.

Test monitor and observers   The test monitor is in charge of the test. This person is responsible for preparing test materials and coordinating all the test-related efforts. The ideal candidate for this role is a usability engineer with a good memory and excellent organizational skills.

Identify individuals to form an observing team. This team could consist of programmers, technical writers, and managers.

Helpers   You need one knowledgeable programmer who can help if the software system being tested gives trouble or crashes.

Have one or two people assigned for logging data and keeping track of the start, end, and elapsed time for tasks. In addition, you will need a video operator and other equipment operators, depending on the equipment you are using in the test. The video operator should capture everything about the test, including all interaction between test participant, software system, and test monitor.

14.4.1.4   Set up lab
See Figure 14.2 for a simple yet useful usability lab. You need two adjacent rooms. Room one is where the software system will be tested by the participants. Install the software and make sure it works fine. Also install any hardware, such as printer, that may be required as part of the test. You might want to have remote-controlled video cameras to capture the progress of the tests. Room two will have the test monitor and observers, watching the test. You could have a one-way mirror between the rooms so that the test could be directly seen by the observers. To allow the test participant and the test monitor to communicate, you will need a two-way audio system.

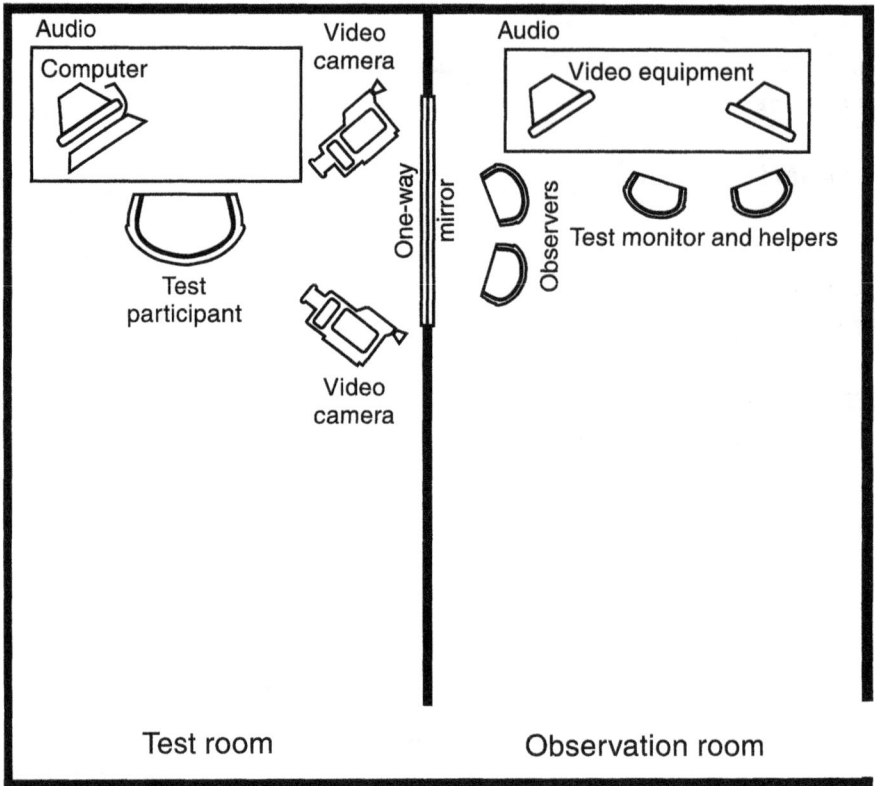

Figure 14.2   A simple usability lab.

Instead of manually logging data, you could use software tools. A data-logging program helps, among other things, log the time the participant spends on completing tasks or reading online Help.

14.4.1.5   Conduct test

Conduct a pilot test. Take the pilot test yourself with help from colleagues. Check and ensure that the equipment and the testing environment are in order.

On the day of the actual test, meet the participants and explain what you expect from them. Remember to assure them that only the software is tested—not the test participants.

Provide the minimum information required for participants to begin. Do not instruct them about solving the task scenarios. Also, do not instruct them about using this or that information element. Provide assistance only when they cannot proceed by themselves. During the test, the technical writer should look closely at how information is used, what problems users have, and what users say.

Collect data. You can use an automatic data logger or you can enter data yourself on the computer, where automation includes letting you quickly enter codes for events, such as START TASK. Enter comments such as, "Does not follow the sequence in the manual." Manual methods use forms (Figure 14.3) for capturing both quantitative and subjective data. Form design depends on what data you need.

At the end of the test, collect the test participant's feedback through interviews.

14.4.1.6 Debrief test participants

Sit with the test participant and review his or her actions during the test. You may want to replay the video to help refresh the participant's memory.

Participant name	Date	Time

Task	Elapsed time	Observations

Figure 14.3 A simple data-collection form.

14.4.1.7   Analyze data and recommend solutions

Replay videotape and find areas where test participants had problems. Note the comments participants made during the test and their answers to the questionnaire. Identify and focus on those tasks that did not meet your goals.

Prepare a report. List all the major problem areas. Offer a solution to each problem area and assign severity codes. Focus on solutions that will have the widest impact on the software system.

14.4.2   Validation testing

Validation tests are ones you should perform to ensure that information is correct and that information elements work correctly. Depending on your test objectives, you can probably perform these tests without lab facilities, without test participants, sometimes even without the software system being developed. Here are some validation tests you should perform:

- Test all procedural steps provided.

- Check if all the examples used in support information and in messages really work.

- Test the context-sensitivity of online Help.

- Check messages alongside corresponding message Help to ensure correct display.

- Test the correctness of the flow of screens in an online tutorial.

14.5   Reviewing information

Reviews may consist of inspections by users, subject matter experts, editors, and marketing specialists. You should select reviewers carefully. Subject matter expert reviewers can be software engineers in the project team or others such as quality assurance specialists who know the software system being developed. They review at all stages of UCID for technical accuracy. User reviewers are prospective users or people who are representative users in terms of various characteristics like educational background and job experience. Their inputs will help you understand the usability of the entire product, including information. Marketing reviewers ensure that information elements project a correct and positive company image and protect confidentiality of corporate information. Editors review for the look and feel of text and graphics. They can often assume the user's role and review from that perspective.

We will look at various types of reviews you should have: technical review, user review, and management or marketing review. Before that, we will start by looking at the review process:

1. Identify reviewers;

2. Provide a good "draft" for review;

3. Brief reviewers;

4. Review "draft";

5. Report errors.

14.5.1 Identify reviewers

To ensure review success, select reviewers carefully. The reviewer should be qualified. For example, if it is a technical review, the person should know enough about the software system to point out inaccurate information. The reviewer should maintain focus on the review goals (software usability and information quality goals), know that trade-offs may be required, and that every comment may not be incorporated.

14.5.2 Provide a good "draft" for review

A draft can be a prototype, a chapter, or an entire information element. Give your best draft for review. Just because there are reviews, do not give a draft that the writer could have improved. If you normally send out a softcopy for review, make sure you send the latest version. If the information is aimed for the print medium, also send a hardcopy to help reviewers get the "look and feel." If you normally send out a hardcopy for review, provide an unmarked printout. In both cases, the draft should be formatted like a final draft. If the format is expensive or hard to produce for review, provide at least two or three pages in the final format.

14.5.3 Brief reviewers

The review inputs expected from each review can be different. Brief each reviewer about what is expected from his or her review. State review procedures and deadlines. If appropriate, provide checklists such as those in the following sections. Discuss these in a formal review orientation meeting.

14.5.4 Do a technical review

This review is responsible for the technical accuracy of information. It is also responsible for completeness from the user's viewpoint. This review ensures that all information elements are consistent with each other and with the rest of the software. Clearly, it should be performed by members of the programming team. Moreover, you should have chapters reviewed by different people if the functions described in the chapters have been implemented by different people. The following is a sample technical review checklist.

- Are all support and interaction information consistent with each other?
- Are text, examples, tables, and figures correct?
- Does any information element describe features that have been removed?
- Are all required concepts explained?
- Is all task-related information described in proper depth?
- Do examples relate to task? Do they work?
- Are all unfamiliar terms (or labels) introduced within a meaningful context?
- Is each term used consistently to mean the same thing?
- Is each concept consistently referred to by the same term?

14.5.5 Do a user review

A review by existing or prospective users is critical for ensuring a successful software system. See below for a sample user review checklist. While implementing review comments, you should consider giving priority to comments received from users.

- Is the information element appropriately organized?

- Is the information element usable?

- Is information retrievable?

- Is information understandable?

- Are examples clear and useful?

- Are figures useful and meaningful? Do they support text?

- Are all unknown terms (or labels) defined?

- Are TOC and index useful?

- Do size and shape of printed element fit purpose?

- Is the technical level appropriate?

14.5.6   Do a management or marketing review

This is a review by a marketing specialist or a management person in your organization. One objective of this review is to ensure that the software system is described as it will be marketed. This takes care of appropriate product positioning. Another objective is to ensure that the presentation of information is in step with the organization's corporate image. Below is a list of more objectives.

- Are the software features presented clearly?

- Is the technical level appropriate?

- Is proprietary and confidential information eliminated?

- Are the right names used?

- Are presentation, style, and terminology consistent with other product communications?

- Are there any trademark violations?

- Is overall presentation neat and attractive?

■ Are size and shape of printed elements suitable for intended users, purpose, and place?

14.6 Editing information

All information should be edited. Two types of editing—technical edit and copy edit—are essential. We are not talking about self-editing, where each writer is responsible for his or her own editing. We are not talking about peer editing either, where writers edit drafts of other writers. Both are useful techniques you can informally use. However, for effective CDG evaluation, you need specialists in editing.

14.6.1 Technical and copy editing

Technical editing is a critical activity. It is not mere spell checking or "correcting a document," as some believe. Good technical editing contributes significantly to critical design goals such as retrievability, readability, and clarity. Moreover, it brings consistency across all information elements. Technical edit goes into content and meaning and is required right from the early stages of the UCID process. It may involve a detailed edit of paragraphs. Technical editors who have come from a technical writing background can also contribute to planning, outlining, and prototyping activities. They can establish standards and guidelines for the organization and for the project, and offer in-house training programs in technical communication.

Copy edit takes care of things like writing style, mechanical style, grammar, and format. As the focus is on the surface structure of the information element, copy edit requires little or no technical knowledge of the product. Unless you have a separate team of copy editors, the copy-editing function should also be performed by the technical editor. Errors in grammar, spelling, and punctuation call attention to themselves and hinder comprehension. For copy editing, you should consider following a standard such as *The Chicago Manual of Style*.

Using checklists and quality indicators, editors should do the following:

1. Edit for retrievability. Check for organization approach and retrieval devices such as an index.

2. Edit for readability. Check page (or screen) layout, formatting, typography, color, and so forth.

3. Edit for clarity. Check for user viewpoint writing, conciseness, coherence, consistency, examples, and so forth.

4. Check for completeness. Particularly ensure that all types of information covered in the prototype and all the topics specified in the outline are covered at the required level of detail.

5. Check for conformance to specifications and prototypes.

6. Edit for appropriate writing style and mechanical style. Ensure conformance to corporate or other agreed upon standards.

7. Check for integrity. Check for page numbers, figures, tables, footnotes, cross-references, chapters, table of contents, and index:

   Within and across chapters;

   Within and across information elements (including labels and messages).

8. Do a print/production check. Ensure the quality of things like figures, cover page design, paper, and binding.

Whether it is a large draft of a printed manual or a handful of labels, editorial review and editing take time. Therefore you need to plan for this effort and keep editors informed well in advance, especially if they are not dedicated full-time to your project. Preferably, editors should be involved in the project right from the start. They should know the standards and guidelines to be followed for the project.

14.6.2 Sample quality indicators

Indicators are measurable characteristics that affect the quality of information. Indicators are helpful for technical writers and editors. With indicators, you know how to improve evolving drafts of an information element. You should, of course, have reliable indicators defined early in the project.

Indicators for retrievability:

- Size of index.

- Size of table of contents.

- Number of headings.

Indicators for readability:

- Reading grade level.

- Number of acronyms and technical terms.

Indicators for clarity:

- Number of examples.

- Number of graphics.

- Average number of steps in tasks.

14.6.3   Sample editing checklists

Checklists are helpful for the person who edits information. In this section, you will find sample checklists that you can use as a starting point.

Retrievability: Sample Checklist

Retrievability is the speed at which users can get the information they want.

- Does the information element's title indicate its users, purpose, and content?

- Is similar information organized and presented in a similar way?

- Is information organized the way users would expect and need?

- Do "parts" consist of related chapters?

- Do chapter titles guide users to required topics?
- Do chapter introductions quickly indicate the content, purpose, and users of the chapter?
- Is the chapter broken down into sections with useful headings?
- Do headings guide users to required topics?
- Are heading levels visually identified?
- Do headings reflect tasks and information to which they relate?
- Do headings show how the parts of an information element fit together?
- Does the TOC show three or four levels of headings?
- Is a figure list included? Do figure titles help identify desired figure?
- Is a table list included? Do table titles help identify desired table?
- Does the index have these entries: tasks, features, topics, new or unfamiliar terms, synonyms, abbreviations and acronyms and their spelled-out versions?
- Are index entries worded for quick retrieval?

Clarity: Sample Checklist

Clear information is that which is easily, quickly, and correctly understood by users.

- Do simple concepts precede more complex concepts?
- Are unnecessary new terms avoided?
- Are approved terms used?
- Are all abbreviations and acronyms expanded in body text?
- Are all unfamiliar terms defined at first use?
- Is each term used consistently to mean the same thing across all information elements?
- Is each concept consistently referred to by the same term?

- Are examples useful? Do they work?

- Are figures helpful?

- Are there enough figures?

Writing Style: Sample Checklist

A good writing style makes things clear to the user. It also makes the reading effort more pleasant.

- Does each paragraph present one main idea? Is there a topic sentence?

- Is each paragraph less than half a page of text?

- Are sentences simple?

- Does active voice predominate?

- Does flow of information promote coherence?

- Are transitions between topics smooth?

- Are instructions in imperative voice?

- Are abbreviations and acronyms used with discretion?

- Are all cross-references necessary?

- Does writing suit reading level of user?

- Have the information elements been spell checked?

Mechanical Style: Sample Checklist

Many editors may not consider editing for mechanics a challenging task, but good mechanical style can often make your information elements look professional.

- Are spelling, capitalization, punctuation, and grammar correct?

- Are all numbered figures referred to in the text?

- Do all references to figures precede figures?

- Are running heads and feet correct?

- Are all cross-references correct?

- Does the page numbering scheme aid retrievability?

Aesthetics and Production: Sample Checklist

To users, good aesthetics can mean pleasant-looking and easy-to-read information elements. Good production can also mean that your information element is easy to use in the product-use environment.

- Are cover and binding strong enough for expected life of information element?

- Is cover design in harmony with the pages of the information element?

- Is page format consistent with standards?

Reference

[1] Rubin, Jeffery, *Handbook of Usability Testing: How to Plan, Design, and Conduct Effective Tests*, New York, NY: John Wiley & Sons, Inc., 1994.

Glossary

ARRCUPA Acronym for these information design goals: availability, retrievability, readability, clarity, usability, pleasantness, and accuracy.

automatic Help Contextual information that the software automatically displays when the cursor is on or moved over a user object such as an entry field. Automatic Help does not require any user action by way of explicit key press or mouse-click.

CDG evaluation The second level in the UCID two-level evaluation approach. CDG evaluation involves the testing, review, and editing of each information element to determine to what extent the element meets the critical design goals (CDGs) earlier defined.

CDG See critical design goals.

critical design goals (CDG) A set of information design goals critical for the success of a specific information element. The goals for information quality are defined based on CDGs.

end-use task   A task that is specific to the software in question. In a software package meant for maintaining employee details, "Entering employee details" is an end-use task. Contrast with support task.

error message   A message that tells users that something has gone wrong. An error message requires users to do something before they can continue with the task they were performing.

feedback   See message.

information architecture   Blueprint for maximizing software usability via the integrated design of all the four information components. The information architecture identifies all the information elements that users need and expect, describing each in terms of content, media, and form.

information components   Software usability components that comprise the various information elements. The four information components are: 1) labels, 2) messages, 3) online support information elements, and 4) printed support information elements. Components 1 and 2 are collectively called interaction information, whereas 3 and 4 are collectively called support information.

information design goals   Quality factors for the effective design of information. The most important factors are covered in the acronym ARRCUPA, which stands for availability, retrievability, readability, clarity, pleasantness, and accuracy.

information design   The design of information elements for effective communication.

information element   One of many separately packaged pieces of information such as field Help, online tutorial, user's guide, and reference card.

information quality goals   User performance and user preference related goals against which information elements are evaluated. Information quality goals are defined based on critical design goals.

information requirements   Information that users need and expect for performing tasks with a software.

information use model   A description of users' requirements concerning the use of an information element. One such requirement could be "Information can be easily and quickly obtained."

information   All the textual elements that software users see and use, including but not limited to messages and printed manuals.

informative message   A message that tells users about what has happened or what is currently happening. An informative message does not require any action for users to continue with the task they were performing.

integration evaluation   The first level in the UCID two-level evaluation approach. Integration evaluation involves the testing of the entire software, covering all four software usability components, including the information components of labels, messages, online support information elements, and printed support information elements.

integration   The software usability driven approach to identifying all the information elements users need and expect, and designing, evaluating, and improving them as an integrated whole.

interaction components   Software usability components that are essential for people to use software. The two interaction components are 1) user objects and 2) user actions.

interaction information   Information that is essential for successful user-software interaction. Labels and messages are interaction information.

label   Text, icon, or combination that identifies a user object.

message   Text, graphic, or combination feedback from software in response to user action or change in system status. The term is used interchangeably with feedback.

prototype   A representative model of 1) a distinct portion of an information element or 2) an entire information element. An example for a prototype could be a chapter of a user's guide. The prototype will have all the items planned to be covered, complete with final-draft-like formatting and figures.

requested Help   Contextual information that the software displays when users press, say, a shortcut key such as F1 provided for that purpose.

retrieved Help   Information that users manually find, select, and display with the help of retrievability devices such as an index.

software usability components   Parts (of a software package) whose quality impacts the ease, speed, and pleasantness with which users can use the software. The software usability components are 1) user objects, 2) user actions, 3) interaction information, and 4) support information.

software usability goal   User performance- and user preference-related goals against which overall usability of the software is evaluated.

software usability   The ease, speed, and pleasantness with which users can perform end-use and support tasks with the software.

specifications   The low-level design details for the information elements defined in the information architecture. For each information element, the specifications document records the purpose and users, content outline, critical design goals, and any production issues.

support information   Information that helps users understand the other software usability components—that is, user objects, user actions, and interaction information (labels and messages). Online information elements such as Help and printed information elements such as a reference card are support information.

support task   A task that is generally required to be completed before users can perform (or effectively perform) end-use tasks. Installation and customization are examples of support tasks.

task description   Details about each major task users can perform with the software. A task description answers questions such as "What is the task?," "With what computing resources is it performed?," and "How do users recover from errors?"

task   A set of actions performed by a software user to achieve a goal. A task can be an end-use task such as "Entering employee details" or a support task such as "Installing the system."

technical writer   A professional who can understand computers, and think and write from the computer user's point of view.

UCID plan   An evolving document that records what information elements should be developed and how.

usability engineer   A professional who understands the strengths and limitations of users, and can design and evaluate software for usability.

usability engineering   The professional practice of engineering a software for usability.

usability plan   An evolving document that records how the software will be designed and evaluated to ensure usability.

user actions   Interactions users have with user objects. An example of a user action is a mouse-click a user performs to select an icon.

user objects   Items users see (and interact with) on the software's screen. Examples are windows, menus, fields, and buttons.

user profile   Details about a group of users with similar characteristics. A user profile covers a user group's general characteristics (learning skills, memory, etc.) and specific characteristics (application domain knowledge, user interface familiarity, etc.).

user   A person who will use the software being developed or someone closely representing the users described in the user profile.

user-centered design   See usability engineering.

user-centered information design (UCID)   The software usability driven approach to designing all four information components via an integrated method. Iterative design and a two-level evaluation approach are followed. Technical writers, in collaboration with usability and software engineers, primarily write all the information components including labels and messages.

warning message   A message that tells users that something undesired might happened if left unattended. No immediate response from users is required for them to continue with their current task.

About the Author

PRADEEP HENRY has over 12 years professional experience in technical communication and usability engineering. He has written and edited for IBM's programming labs worldwide and has designed and taught corporate courses in technical writing and user interface design. He pioneered the quality technical communication culture in India and was profiled as Distinguished Professional Communicator in *IEEE Professional Communication Newsletter* (March/April 1994).

In his current employment, Pradeep Henry leads a group responsible for technical communication, usability engineering, and website design. He is a member of the ACM Special Interest Group for Computer Human Interaction and the IEEE Professional Communication Society. Having received a B.Sc. in physics and an M.A. in English from the University of Madras, he has also recently taken a software usability engineering course at University of California at Berkeley (Extension).

Index

The Artech House Computer Science Library